*B*ird **T**raditions
of the
Lime Village Area
Dena'ina

Upper Stony River
Ethno-Ornithology

Bird Traditions
of the
Lime Village Area
Dena'ina

Upper Stony River
Ethno-Ornithology

Priscilla N. Russell
George C. West

with editorial and linguistic comments by James Kari
illustrations by George C. West

Alaska Native Knowledge Network

Alaska Native Knowledge Network
Center for Cross-Cultural Studies, University of Alaska Fairbanks
PO Box 756730
Fairbanks, Alaska 99775-6730
www.ankn.uaf.edu

Elmer E. Rasmuson Library Cataloging in Publication Data:

Russell, Priscilla N.
    Bird traditions of the Lime Village area Dena'ina / Priscilla N.
Russell, George C. West ; with editorial and linguistic comments by
James Kari ; illustrations by George C. West.—Fairbanks : Alaska
Native Knowledge Network, Center for Cross-Cultural Studies,
University of Alaska Fairbanks, 2003.

    p. : ill. ; cm.
    Includes bibliographical references.
    ISBN 1-877962-38-4

    1. Dena'ina Indians—Ethnozoology—Alaska—Lime Village.
2. Ornithology–Alaska—Lime Village. 3. Dena'ina language—
Alaska—Lime Village. 4. Athapascan Indians—Alaska—Lime
Village.
    I. Title. II. Russell, Priscilla N. III. West, George C.
IV. Kari, James M.

    E99.T185 R87 2003

This material is based, in part, upon work supported by the National Endowment
for the Humanities and the National Science Foundation. Any opinions, findings,
and conclusions or recommendations expressed in this material are those of the
authors and do not necessarily reflect the views of the above agencies.

COVER PHOTOS: Front cover photo of the late Vonga Bobby, Lime Village
traditional chief for many years. Back cover photos of Pete Bobby and Emma Alexie.

Dedicated to the Lime Village Elders, especially the late Vonga Bobby, a true Dena'ina chief, and his wife Matrona whose gracious hospitality inspired this report and also to Pete Bobby, a great teacher. This report is possible only because of their commitment to their cultural heritage that they demonstrate by actions and words. By the elders sharing their knowledge, the younger generations have learned and the culture continues to live. The word "elder" can have more than one meaning. Primarily we use it to refer to the oldest generations. Secondly, we use it to refer to any person who teaches a younger person.

Contents

Photo insert following page 46

Classification of Birds 48

Appendices

Foreword

This book on the use of and relationship with birds of the upper Stony River Dena'ina is divided into two major sections. The first section contains an introduction to the upper Stony River Dena'ina and their country. Among other subjects, it describes their harvesting strategies, uses, beliefs, and social interactions relating to birds. The second section is comprised of species accounts with descriptions of the bird's physical appearance and habitat and often its vocalizations and habits. Most accounts include ethnographic information about the species. The second section is organized to reflect as much as possible the Dena'ina bird classification system.

Although the majority of the information was obtained through informal interviews and field observations between the fall of 1991 and the spring of 1992, some data were gathered during other visits to Lime Village in the 1970s and 1980s. Field guide photographs and drawings aided the identification of some birds. The second author assisted in formulating questions to aid in determining which species were in the area and to elicit specific information about certain species. He is responsible for the descriptions of species in the second section, for editing the text, and for the illustrations in the text. The primary author collected the ethnographic information and is accountable for any errors in the ethnographic material and interpretations.

The primary author chose to use the present perfect tense as the main tense in the ethnographic portion of the paper because she did not feel she had sufficient knowledge to always record accurately which traditions belong solely to the past and which ones have continued into modern times. For the purposes of this report, modern times are defined as the last thirty years. Consultants also indicated feeling more comfortable with an indefinite tense when the primary author gave them an early draft to review.

The majority of information was shared by Lime Village residents. We chose to refer to the people as Lime Village area Dena'ina for several reasons. Lime Village elders, who contributed a significant portion of the information, were semi-nomadic in their early years and ranged large distances from what is now Lime Village. Although, perhaps to a lesser extent, this is also true of younger generations. For various reasons, this ethno-ornithology is not a complete record of bird uses in the Lime Village Dena'ina area.

The first published source on Dena'ina bird names for several dialects is Osgood (1937, 209–210). The list of Dena'ina bird names in the Lime Village area has been advanced and refined since the 1970s. J. Kari (1974, 9–10) has a list of bird names for the Kenai Dena'ina dialect. Tenenbaum (1975, 12–17) has a very good bird list for Nondalton with much information provided by Pete Trefon, Antone Evan, and Macy Hobson. These sources were combined and advanced in the Dena'ina bird list in J. Kari (1977, 42–53). P.R. Kari (1983, 136–37) has a list of the birds used for subsistence in the Lime Village area. All of these sources were used by Russell in her work in Lime in (1991 and 1992). The bird list in J. Kari (1994, 19–27) incorporated Russell's later research. This book has the most current information on bird names and identifications for the Lime Village area including the comments of George West.

The order in which species appear in the second section is discussed under "Classification." Only those species verified as occurring in the upper Stony River area are listed in the first nine sections of the species accounts. In those sections we also include several birds that are recognized by Lime Village people but for which there is no Dena'ina name. The tenth section contains "Probable Species" that should be in the Lime Village area but were not identified by the primary author or the Dena'ina of the area, and several species that have been reported to be there but could not be verified (West 2002).

The format for the second section is as follows: If a Dena'ina name is known for a major group, that name follows the English group name. For each species, the English common name is given followed by the scientific name in parenthesis, followed by the local common name, if any, in quotation marks. Below the common name is the Dena'ina name printed in italics to help distinguish it from English words. If the Dena'ina name is not known, a question mark appears (?). A question mark is also used if we are unsure of the Dena'ina name or if the Dena'ina name is incomplete. After the Dena'ina name is the translation of the name, presented in quotes. If the Dena'ina name is a stem word, the symbol "√" is given. A stem word is usually one syllable and cannot be broken into smaller parts. If part of the translation of the Dena'ina name has not been determined, "?" will be noted in the translation line.

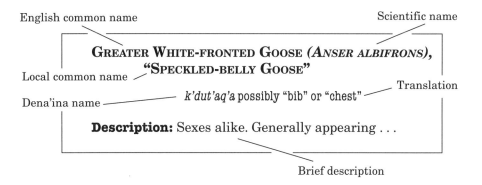

English common name

Scientific name

GREATER WHITE-FRONTED GOOSE (*ANSER ALBIFRONS*), "SPECKLED-BELLY GOOSE"

Local common name

k'dut'aq'a possibly "bib" or "chest"

Translation

Dena'ina name

Description: Sexes alike. Generally appearing . . .

Brief description

A brief description of each species is given along with voice, behavior, and habitat in order to help to identify the bird. Ethnographic information follows the description.

Because the Dena'ina people have made a gift of their knowledge to this book, the authors have declined any proceeds from its sale.

Acknowledgments

As primary author I give my very, very special thanks to the residents of Lime Village who made this report possible. I decided not to list the individual names of all the consultants because the majority of residents, from the elders to the children, shared their knowledge. I also thank the Lime Village people very much for their gracious hospitality and for providing other essential needs.

The vast majority of the cultural information in this paper was shared by present-day Lime Village residents. However, Dena'ina with strong connections to the area who live elsewhere also contributed to it, for which I thank them.

I would be remiss in not acknowledging my coauthor, George West. The quality of this book is greatly enhanced due to his expertise and many extra hours of careful work.

I warmly thank Melinda Moore, former Lime Village public school teacher, for initiating a formal bird study as part of the curriculum and the students for teaching me (see Appendix A). I also thank Phil Graham, former Lime Village public school teacher, for his assistance through the school system.

I am grateful to the late Luthur Hobson Sr. for acting as a translator and to James A. Fall, Division of Subsistence, Alaska Department of Fish and Game, Anchorage for serving as an evaluator for this project and the report.

As coauthors, we also thank Brian McCaffery, Wildlife Biologist, U.S. Fish and Wildlife Service, Bethel, Alaska for his valuable information on the probable occurrence of birds in the upper Stony River area. We also thank Paula Elmes of the Alaska Native Knowledge Network for her help with the final layout. We thank Arthur Kruski for proof reading a portion of the text.

We sincerely thank the Homer Society of Natural History's Pratt Museum for funding a portion of this work and especially Betsy Pitzman and Patti Carey for providing their invaluable support and assistance.

This project was supported in part by a grant in 1991 from the Alaska Humanities Forum and the National Endowment for the Humanities, with additional publishing assistance provided through the Alaska Native Knowledge Network, to whom we extend our deep appreciation.

Our great hope is that this publication will remind Dena'ina people of all ages that they can be truly proud of their unique cultural heritage and that the traditions will continue.

Priscilla Russell
George West

Introduction

L ime Village is a remote Dena'ina community located on the
lower portion of the upper Stony River in the western foothills of the Alaska Range. It is situated about 180 air miles
west of Anchorage and approximately 60 air miles and 75 river
miles upstream from where the Stony River joins the Kuskokwim
River (see map on pages 4 and 5.) For the purposes of this paper
the upper Stony River begins in the Tishimna Lake area approximately 30 miles west of Lime Village.

According to the 1990 census figures, Lime Village is composed of
42 people (Bureau of the Census 1992, 42). Although intermarriage
has occurred with other ethnic groups, Dena'ina Athabascan continues to be the dominant traditional heritage of the people.

At one time Lime Village Dena'ina elders and their ancestors were
semi-nomadic not only along the Stony River drainage but, among
other places, the upper Swift River and Telequana Lake areas. Besides the Stony River area, Lime Village residents have continued
to use other traditional areas in modern times. See Kari and Kari
(1982), P. R. Kari (1983), and J. Kari (1988) for a more complete
discussion of the Inland Dena'ina land use area.

Environment

Environmental Communities

The major environmental systems of the upper Stony River area are briefly described below. They are primarily taken from Viereck and Little (1972, 14–23) with modifications by Viereck and Dyrness (1980, 1–38) and David Murray (personal communication).

Closed Spruce-Hardwood Forest: This relatively tall, dense forest system of low to middle elevations grows on moderately to well-drained locations. Cottonwood *(Populus balsamifera)* dominates the floodplains while paper birch *(Betula papyrifera)* and white spruce *(Picea alba)* are abundant at higher elevations.

Open, Low-growing Spruce Forest: The dominant black spruce *(Picea mariana)* of this poorly drained system may be mixed with paper birch, tamarack *(Larix laricina)*, and treeless bogs.

Bog: This very wet, treeless environment contains shrubs, ponds, and other areas of standing water.

Shrub Thicket: Alders *(Alnus* sp.) above tree line and willows *(Salix* sp.) along streams dominate this largely high brush system. It exists in a variety of situations including between tree line and alpine tundra, between forest and beach, on floodplains, and in avalanche paths and other disturbed areas.

Moist Tundra: Small shrubs and lichens are the dominant vegetation in this treeless environment that is prevalent in the foothills and lower elevations of the Alaska Range.

Alpine Tundra: Occurring over 2,500 feet in the mountains above the tree and brush zones, this dry environment is characterized by small herbaceous and woody plants. Lichens also prevail.

Aquatic Environments: Lakes, ponds, streams of varying sizes, and wetlands are abundant in the upper Stony River area.

Glaciers: These areas of ice and snow are common in the Alaska Range at head of the Stony River on the west slope of the Southern Alaska Range.

Climate and Weather

The upper Stony River area is affected by a mountainous transitional climate close to and within the Alaska Range and a continental climate elsewhere. Since Lime Village receives weather from both of these systems, weather in the area is variable. Cool to warm summers with considerable rain and overcast days are normal. Although freezing night temperatures may occur in late August, freeze-up usually happens between mid-October and mid-November. Below freezing winter temperatures are normal with very cold periods of −30° to −50° F common. Lengthening daylight hours that cause temperatures to rise above freezing for a portion of sunny days especially and to fall below freezing at night begins in March and may continue through most of April. Breakup caused by temperatures frequently above freezing during the day and often at night normally starts in April and continues well into May. At this time, traveling conditions are very poor due to much excess surface water and wet, slushy snow and ice.

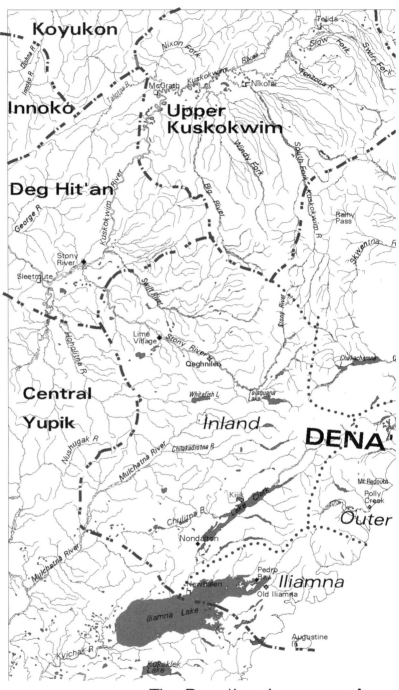

The Dena'ina Language Area

Language
names &
boundaries

Dialect
names &
boundaries

4

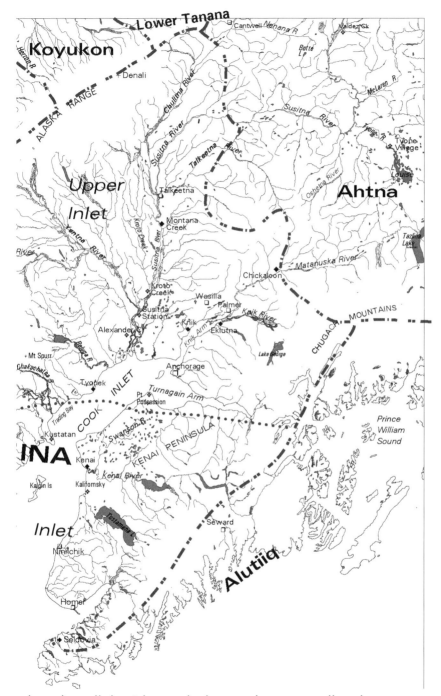

showing dialect boundaries and surrounding languages

◆ Modern settlements with Dena'ina population
❖ Some former Dena'ina settlements of the historic period (now abandoned)
▢ Other modern towns and villages

Seasonal Cycle

The Lime Village bird year is divided into four main seasons: winter, spring, summer, and fall. The following divisions are approximate because weather and other environmental conditions, upon which the seasons are largely based, influence bird migrations. Winter is the longest season, extending from mid-October or early November until March or early April when the first migrating birds arrive to signal the advent of spring. Summer begins in early June when the last of the area's migrants, known locally as "summer birds," return. Fall is signaled by the southerly migration of large flocks, beginning as early as late August and continuing into October. Birds that are permanent residents or remain throughout at least a portion of the winter are referred to locally as "winter birds." Examples of permanent winter birds are chickadees, gray jays, ravens, and pine grosbeaks while nomadic (wandering) winter birds include redpolls, Bohemian waxwings, and pine siskins.

Traditionally the Lime Village Dena'ina believe that migrating birds fly to a warmer, salt-water environment when the weather becomes too cold for them in the Interior and come back from there in the spring. They ask the birds to return when they leave, and they tell them that they are happy to see them when they reappear. The spring return of migrating birds is an especially joyous occasion. People greatly appreciate the variety of life that the new arrivals add with their songs, beauty, and activity. The fresh food provided by some species has provided a welcome change from the winter diet and at times served as survival food.

Eagles and hawks return in March and April either before or at approximately the same time the first waterbirds arrive. On the other hand, during a warm winter some of these birds of prey may not migrate. (The goshawk is a year round inhabitant of the area.) The Lime Village Dena'ina name for March, *ndałika'a n'u*, translates as "bald eagle month" indicating the return of this bird at that time. April is called *q'uluq'eya n'u* which means "soaring hawk month" (J. Kari 1977, 144).

Lime Village people have hunted waterbirds primarily in the spring and fall when the birds are returning or leaving or migrating through the area. They have begun hunting waterfowl in late March or early April when the first birds arrive and have continued to hunt them throughout the spring until the birds nest. People have resumed harvesting waterfowl in August and September when molting is complete and the young are independent. Cranes, most shorebirds, and gulls arrive in early May after the majority of the waterfowl. While most shorebirds leave by early fall, gulls may stay until water bodies freeze over.

Snow buntings are the first small birds to migrate through the upper Stony River area between late March and early April. An elder observes that they appear around Russian Orthodox Easter. In earlier times, especially when there was a food shortage, the Dena'ina hunted snow buntings. The Lapland longspur is another small, non-resident, early spring migrant seen in open country and near lakes where waterfowl rest, feed, and nest.

Early forest song birds that normally appear in April and reside in the area are the ruby-crowned kinglet, varied thrush, and dark-eyed junco. They are followed by the robin and eventually the other thrushes. Most other small summer nesting birds arrive in May or early June.

Grouse, permanent resident game birds, have been hunted throughout the year except in the spring and early summer when the birds are nesting and raising young. Ptarmigan, largely upland resident game birds, have been harvested primarily between October and April.

Social
Interaction

Learning about Birds

In a traditional Dena'ina society, education through observation and practice is at least as important as verbal instruction. Listening to traditional stories and personal accounts about birds is another valuable way of acquiring knowledge. Learning about birds involves all these methods.

BIRD IDENTIFICATION

People identify birds by their songs and calls, physical appearance, habitat, and habits. Sound appears to have been an especially important way of identifying birds before binoculars were introduced, a method that has not lost its importance. A significant number of Dena'ina bird names imitate or are based on bird sounds. See discussion in "Communication With Birds."

The Lime Village Dena'ina are expert hunters and fisherman and consequently they have an intimate knowledge of the anatomy of all animals in their country. Bird anatomy makes use of general anatomical terms and some unique bird anatomical terms.

LEARNING BEFORE HUNTING

Young children have accompanied adults, often their grandparents,

Dena'ina Terms for Parts of a Bird

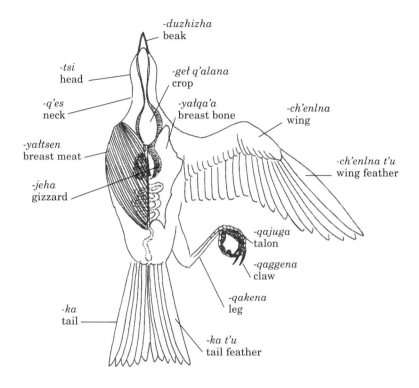

who identify birds for them while teaching them their names and characteristics. When old enough, they have also learned about birds by exploring alone and with their friends. For example, they have found bird nests and observed the related activity. Children have live-trapped birds, watching them for a short period of time, and then releasing them. They have learned to do bird-related chores such as plucking birds by first observing the work being done and eventually helping with it. Plucking gives a child an excellent opportunity to closely observe a bird.

LEARNING TO HUNT

Before going on an extended hunt, children have been taught weapon safety and allowed supervised practice near home or camp. Slingshots and "BB" guns have often been their first weapons and gray jays and other nuisance birds their first prey. Frequently grouse is the first game bird of a young hunter.

A teacher is said to instinctively know when a child is ready to hunt and contribute to the hunt. A primary way that children have learned to hunt is by accompanying a skilled hunter and watching him hunt. When the child appears ready to use a weapon in a real hunting situation, he is given the opportunity. Although exceptions exist, traditionally uncles have taught nephews their primary hunting skills. The following examples illustrate a variety of situations.

> A young adult male recalls that between the ages of eight and thirteen, he accompanied his older cousin duck hunting. By watching his cousin closely, the boy was able to shoot a duck when first given the opportunity.

> A man relates that his grandfather told him how to shoot a gun when he was eight years old but that his mother showed him how when he asked her. He accompanied both his grandmother and uncle on hunting trips and credits them for teaching him hunting skills.

> An elderly woman remembers that as a child she often hunted with her father. A male elder largely learned from his father and grandfather. Since then, he has taught many younger people, both girls and boys, to hunt by verbally instructing them, showing them, and allowing them to participate in activities. For example, while he makes a blind, his students may both watch him and bring materials.

Sharing the Catch

To show respect, at least a portion of the first waterfowl catch of the season has ideally been given to the elders of the community. The return of the birds and their first waterfowl meal is said to be an especially happy event for elders and a sign that they survived another winter. The remaining birds are distributed throughout the community so that everyone can participate in the joyous occasion.

Traditionally waterfowl have been widely shared among members at camp. Although a delicacy, waterfowl have served as emergency food especially early in the spring season when other food may be in short supply and traveling conditions very difficult. When enough waterbirds have been harvested, hunters return to the village and distribute them. The way birds are shared may depend on a variety of factors. After the initial harvests, typically a hunter gives his catch to members of his household and other close kin. If he has harvested a large number of birds, he is more likely to further distribute them. Hunters also donate large catches to funeral potlatches and other ceremonial events.

Harvesting Strategies

The Harvesting of Birds

Traditionally the upper Stony River people have used two primary strategies for harvesting birds. In one strategy, the hunter sets stationary objects such as snares, deadfalls, and nets. Although he checks them regularly, he does not have to be present to catch the bird. In the second strategy, the hunter uses bows and arrows, throwing weapons, and in more recent times, guns. The employment of those weapons demands the active involvement of the hunter in the bird's death. Following is a description of the two harvesting strategies and associated equipment.

SNARING

Snaring has been one of the most common traditional methods of harvesting birds. An elder comments that generally it is the most humane way because usually the bird is snared by the neck and quickly dies. Although other birds including cranes and eagles have been snared, waterfowl, grouse, and ptarmigan appear to have been the most frequently snared birds. Snares have been used in the majority of environments that birds inhabit and a variety of techniques have been developed for employing them.

The most common snaring method that can be used for any type of bird is to simply set individual snares where birds tend to rest, land, feed, or be otherwise active. The snares are hung on poles or branches at the height of the bird's neck so that the bird is hung by its neck and dies quickly. Snares are placed on ice where waterfowl are known to land, on banks and beaches where they walk to and from the water, on beach logs where they rest, and in vegetation along shores where they feed. Eagle snares have been located in the bird's habitual landing spots and crane snares are placed where they feed.

People have piled willow and other shrubs in approximately two foot high horizontal rows or "brush fences" (called *heł* in Dena'ina) and placed snares in the brush to capture ptarmigan and grouse. Snares attached to a standing stick at the height of the bird's neck are placed in openings in the fence. An alternative style is to bend the branches holding the snares so that the birds are snared by the feet. In order to remove the birds quickly and prevent unnecessary suffering, the snares are closely watched. Reportedly, waterbirds have also been snared in "brush fences."

To obtain waterfowl, people have stretched lines with fastened snares above small streams and small inlets where birds tend to land. The birds fly or swim into the snares. Traditionally feather or spruce root snares and spruce root or sinew lines have been employed. Spruce root snares are braided with three or four relatively thin roots. Thick roots, difficult to bend and braid, are not used because braiding gives the snare essential strength. Spruce root snares and other snares that tend to become brittle and dry are oiled unless they are placed near water. Although feather shaft snares become dry, they do not break and thus do not need oiling.

To snare waterfowl on lakes, people have constructed small, hand-built rafts. Mud added to the raft holds up the brush to which the baited snares are fastened. The bird is snared by the head and dies on the raft.

Another means of snaring waterfowl on rafts is to tie a snare to a line attached to a rock that holds the trigger in place. When the bird steps on the trigger, it is caught by its feet and pulled by the rock into the water where it drowns. Because the bird is tied to the line, it is easily retrieved.

Besides spruce root snares as described earlier, snares for birds and other wildlife have been made from feather shafts (see Feather Technology). Snares for capturing large birds including ducks have been constructed from dried moose and caribou hide and sinew. Snares placed in wet conditions are waterproofed with spruce pitch. An elder observes that if a person is hungry and has no snares, he can remove the lines from his snowshoes to make snares and replace them when game has been caught. Lime Villagers have constructed snares from bird leg tendons, especially those of larger birds.

NETS

Nets have been stretched across streams for the same purpose as snare lines and in the same manner except that nets may be placed both above and below water. Another traditional means of netting waterfowl is to stretch a trout net between two sticks on a beach. The birds fly or walk into the net which may be baited with water vegetation or other food.

DEADFALLS

Besides snaring birds, deadfalls have been another traditional stationary means of harvesting birds. Usually made of wood and rocks, various types of deadfalls have been used depending on the size and kind of bird to be harvested. Deadfalls for large birds are constructed of logs and rocks. As is true of snares, deadfalls are located in areas that the birds regularly inhabit. For example, waterfowl and crane deadfalls have often been built on beaches and geese deadfalls also are made in shore vegetation where they feed. Waterfowl deadfalls made of wood have also been placed on floating rafts.

ACTIVE HUNTING

In active hunting situations, Lime Villagers explain that the number of needed birds, the humane treatment of the prey, and the safety of people are the most important considerations. For example, it is essential to know where both hunters and non-hunters in the area are located. Before using a weapon, the hunter judges the bird's angle, distance, speed, and the bird's awareness of the hunter's presence. Besides wasting ammunition, misjudgment may wound the prey or spoil the meat if hit at too close range. In some situations, eye contact helps closely located partners know each other's target.

TABLE 1. TERMS FOR DEVICES USED TO HARVEST BIRDS: STATIONARY DEVICES

game lookout	*ninahq'a*	'Vision place'
camoflaged hunting spot, hunting blind	*k'uq'a*	'Opposing place'
elevated rack, platform	*dehq'a*	
small game snare, bird snare	*quggił*	
bird wing feather snare	*k'ts'enlu bila* (Upper Cook Inlet dialect)	
snare barricade fence	*heł*	√
snare lock	*vesexa, vezexa, k'enzexa*	
snare loop	*k'eniq'*	
snare material	*quggił lahi*	
snare sack	*quggił yes*	
floating snare for waterfowl	*quggił hnesa*	'snare raft'
deadfall trap	*ał*	√
floating small deadfall for waterfowl	*ał hnesa*	'deadfall raft'
net	*tahvił*	'water snare'
gunny sack net	*chida yiztl'ini tahvił*	'net woven by an old lady'

Harvesting times are ideally in the early morning and early evening when the birds tend to feed. Traditionally hunters have preferred to be upwind from the birds with the sun at their backs.

WEAPONS

The upper Stony River Dena'ina have used a variety of traditional weapons that involve the active participation of the hunter in the bird's death. These include shooting and throwing weapons that have, for the most part, been replaced by guns. While some types of weapons had a variety of styles, only styles employed in harvesting birds are discussed below.

Besides the standard sharp-pointed arrow, a blunt arrow and an arrow with a sharp, detachable tip have been used. The tip of the latter type, employed primarily for waterbirds, comes loose when it hits its target but remains attached to the shaft by a line. The purpose of the blunt arrow used only on ducks, grouse, and other smaller birds is to knock the bird unconscious or quickly kill it without tearing the body. Because larger birds such as geese and swans are only wounded by the arrow and caused to suffer, it has not been used to hunt them. Waterfowl and other game birds have also been shot with small, sharp arrows that kill them when hit in the head or neck. While shooting a duck or other similar-sized or smaller bird in the body may kill it, hitting a swan or goose on its wings, whether with a gun or arrow, often only wounds it. Because of this, swans especially have normally been snared or killed with a gun but not shot with a bow and arrow.

The best wood for arrows used in wet conditions has apparently been the hard dark wood often located on the windward side of a spruce or on spruce growing in especially cold, wet conditions (P. R. Kari 1987, 28, 29). The arrows are waterproofed with spruce pitch mixed with the right amount of grease (see ibid., 32 for more information). Arrows for hunting land birds have been constructed of birch and not waterproofed with pitch (ibid., 43). Hunting bows have also been crafted primarily from birch.

An elder observes that bow and arrows were largely replaced by .22 caliber rifles and other guns in the early twentieth century. He notes that his grandfather always used a bow and arrow while his father learned to shoot a gun. In modern times, waterfowl and other game birds are usually taken with .22 caliber rifles or shotguns. As was true in the past, hunters have continued to aim at the bird's head, in order to kill the bird quickly without unnecessary pain.

The use of slingshots for killing ducks, grouse, and other small game has continued into modern days. Several people describe making a slingshot from a piece of forked willow because willow is both strong and flexible. The willow ends are tied together tightly with a leather line so that the rock fits snugly between them. A well made sling-shot can hit a fairly distant target, such as a duck on the opposite side of the Stony River.

Besides shooting weapons, the upper Stony River people have made throwing weapons for harvesting birds. One kind is a throwing stick made from a long, stiff stick carved at one end to hold a flat rock. The normal size of stick, which can be made from any kind of wood, is approximately one inch in diameter and three to four feet long. The length of the stick depends on the strength and size of the thrower because the longer the stick, the further the rock can be thrown. The weapon has been used to kill waterfowl and cranes but not grouse and ptarmigan because they can be harvested by a hand-thrown rock.

A sling-rock thrower for killing waterfowl is constructed by tying any shaped rock to one end of a skin line and whirling it to gain speed before sending it to its target. The longer the line, the further the rock travels and the more dangerous the weapon is to the user because it is more likely to hit him as he whirls it. People with strong arms have sent rocks across the Stony River. At least one elder remembers having used a sling rock thrower.

Although weapon technology has changed, many of the same or similar hunting strategies continue to be used in modern times. Because waterfowl have been harvested in greater numbers than other types of birds and more strategies have been developed for hunting them, waterfowl hunting is emphasized. Unless noted, waterfowl have been obtained in the following ways (and other kinds of birds when specifically mentioned).

A common method of hunting with a bow and arrow or gun has been to hide in or behind natural or constructed blinds and wait for an opportunity to shoot a bird. A similar way is to sneak from blind to blind, usually natural blinds, until the right harvesting situation occurs. Surprise hunting gives the hunter a better choice of which bird to shoot. Hiding along a stream or lake bend gives a hunter additional surprise advantage. The bird tends to fall near the hunter when shot at a bend.

Especially during the early part of the season, hunters have waited at stream mouths and areas of ice overflow because waterfowl tend to congregate at these first open water spots. The shallow water on the ice allows them to be easily shot. When ice frozen to the bottom floats up, it brings food with it and thus more birds to the area.

After taking a first shot, the hunter has often remained very still in the same position in hopes of getting another good shot because the birds may circle back or come out of hiding and give him the opportunity to try again. For example, if the bird's mate has been shot, it may return looking for its partner. Unless the meat is not needed, often the mate of a harvested bird is shot to prevent the bird's emotional suffering.

Lime Villagers have harvested birds by standing in an open place and shooting birds flying overhead. Hunters shooting over land have attempted to aim at birds that are very likely to fall near them. If the bird does not land in the vicinity, they make a reasonable effort to find the bird. A child may be sent to retrieve birds on land so that hunters may continue to harvest birds.

Because hunters know the route waterfowl take between lakes, they may stand along the route and shoot as the birds fly above them, or they may wait near where the birds habitually land. They have tended to avoid long shots at overhead birds because they are more difficult to hit accurately and may only wound them. Hard to obtain ammunition may also be wasted. For the same reasons, Lime Village hunters follow this practice in any potentially inaccurate shooting situation.

Regardless of where the hunter is situated, if he hears shots at a different location, he watches carefully in the event that frightened birds may fly his way.

In the early part of the spring waterbird hunting season, a bird that has been shot may land on unsafe ice. One way of retrieving a bird from unsafe ice has been by throwing a line with an attached hook at the bird and snagging it under the wing. Hooks have been made from willow or other available brush. The added weight of the wooden hook makes a more efficient throwing line. In the same type of situation, long poles have also been employed.

Hunters may obtain birds from unstable ice by pushing two logs tied together ahead of them on the ice. If the ice breaks, they hold on to the floating logs until they reach safety. Small watercraft have been used in the same way.

TABLE 2. TERMS FOR DEVICES USED TO HARVEST BIRDS: PORTABLE DEVICES

gun, rifle (any)	*izin*	
arrow	*k'qes* *teldexi*	
blunt arrow for small game; blunt arrowhead	*tl'es*	√
children's arrow	*ch'q'ayna teldexa*	
double-pronged arrow	*duk'nighezhi*	
arrowhead, spearhead	*k'q'u*	√
large game arrowheads	*k'ghaditin,* *k'ghadiliy*	
small game arrowhead	*vekidiluy*	
bow	*ts'iłten*	'single handle'
bow and arrow	*ts'iłten teldexi ts'iłq'u,* *ts'iłten k'qes ts'iłq'u*	
throwing spear, harpoon	*telqexi, qeytełqexi*	
fish spear, beaver spear, harpoon	*tuqesi*	
club, throwing club	*heł*	√
sling rock thrower	*viq' tsateldełi*	
whistle, calling device	*viq' yidelyishi*	

People have tried to shoot near but not directly at waterfowl resting on unsafe ice so as to cause them to fly. They then fall when hit in a more easily retrievable area. For example, people observe that waterfowl like to sun themselves on ice especially during warm afternoons.

Once the ice melts enough to allow hunting from boats, birds have been retrieved from the water as quickly as possible. If no boat is available to reach a bird near shore, rocks may be thrown at it to produce waves that cause it to float to shore or it may be snagged by a pole. Wounded birds, which may dive repeatedly, have normally been followed until they are recovered. In shallow water where boats are not able to travel, hunters have followed a bird on foot while splashing a pole to chase it towards other hunters that harvest it.

To lure birds closer, dead birds have been propped up with a stick as decoys on ice or left floating in the water. The decoys are not left long because they may be preyed upon. If eagles and other birds of prey attempt to take birds that hunters desire, the birds of prey are not shot. Ravens, which are considered scavengers, have been shot near harvested birds to frighten predators.

Regardless of the method used, hunters continue the tradition of being very careful to shoot where the bird can be recovered as easily as possible. They do this both for efficiency and to prevent waste. They also spend the necessary time to reclaim wounded birds and to prevent suffering. A hunter knows how to twist a wounded bird's neck so that the bird dies quickly.

Although men continue to be the primary hunters, the weapons and strategies described here have been used by both women and men who possess the capability.

The Harvesting of Eggs

Lime Village people have collected a variety of wild bird eggs, called *k'ghazha*, for food. The most common kinds have included waterfowl, crane, gull, ptarmigan, grouse, and large shorebird eggs. Customarily waterfowl eggs have been gathered in greater numbers than eggs of other birds. Not only do these waterfowl produce relatively large eggs, but they are laid in great numbers in the spring when food may be very scarce. People explain that elders told them not to bother bird nests unless for food. An elder remembers that swan eggs were only harvested when other food was lacking. Some eggs such as loon and arctic tern eggs normally have not been harvested because of their strong taste. Among other possible reasons such as cultural taboos, small birds eggs have not been regularly gathered for food because of their size. However, all bird eggs may serve as emergency food.

Traditionally fewer female birds were harvested than males. A number of females of each species have been harvested because they tend to be fatter than the males and thus more highly preferred for food. A second essential reason is that the development of the yolk within the female indicates that the females of that species will soon lay eggs and that hunting of the species should end for the spring and summer seasons. Ideally this check is made for each species. Harvested females are boiled and consumed with the undeveloped eggs. It was taboo for a first menstruating girl to eat undeveloped eggs.

Eggs have been gathered soon after they have been laid. People have not normally eaten embryo-developed eggs. Because an egg's outer appearance apparently does not indicate the stage of embryo development, the practice of observing growth within the female bird is necessary for judging appropriate egg harvesting times. If eggs are gathered early in the season, some birds may lay either additional eggs or another clutch.

People have often found eggs by watching the flight of birds to and from their nests. They may remember the location of nests from past years or may accidentally find them. The eggs from only a limited number of each species are taken so as not to deplete the eggs of any one species. If the eggs are taken early in the nesting season, certain species may make another nest. Some people say that one should take all the eggs from a nest of some species because the parent(s) will not return if the nest has been touched.

Traditionally people have boiled eggs or fried them on hot rocks. Eggs have apparently not been preserved for later use.

Composition of Hunting Parties

Although young and middle-aged men have traditionally been the most active hunters, older men, women, and children also hunt game birds. In addition to the relationships described in "Learning to Hunt," a variety of other bird hunting relationships exist. Commonly spouses hunt together as do males or females of similar or varying ages. A person may hunt alone or be one of a party of two to six or occasionally more people.

Transportation

Most waterfowl return in April and May during breakup when human land travel is usually very difficult. At first, snow and ice cover the area except for stream outlets and other select spots such as places with good sun exposure. As the season progresses, melting increases, and this changes the quality and amount of ice, snow, open ground, and open water. These changes not only have demanded varied modes of transportation but alter their use.

With the arrival of the first birds, people not already camped at the hunting grounds have traveled there on snowshoes, dog teams, and in more recent times, snowmachines. Before the waterways have opened sufficiently for boat travel, people have hunted birds on foot. Once enough ice melts, people have also hunted with boats. If necessary, they have slid boats over snow and ice to open water. Heavy watercraft such as aluminum and wooden boats have been pulled by dogs and snowmachines. Although the ice melts first along lake shores, the wind may push it back and forth across the lake producing variable travel situations. By approximately early June, the lake and stream ice has completely melted.

Traditional watercraft include birchbark, moose skin, and caribou skin canoes. Two-holed black bear skin and three-holed brown bear skin kayaks, which could maneuver rapids, were also used for bird hunting. In modern times, aluminum and wooden boats are almost exclusively employed. Canvas canoes were made by local people in the early 1970s for waterfowl hunting.

Because watercraft have not normally been kept at waterfowl harvest sites over winter, in the spring people have transported them overland to the areas. Skin canoes have traditionally been carried by two people and pulled by dog sled. They have been pushed on a well-packed snow trail or in the early morning on firm snow. In lake country, the same means have been used to portage them between water bodies and over ice. In modern times, aluminum and wooden boats have been dragged by dogs and snowmachines to the harvest area or, once it is navigable, to the stream Hek'dichen (Hungry Creek) that flows into the Stony River from the lake area to just above Lime Village. In shallow areas, the boats are frequently poled, pulled, and pushed by hand upstream to prevent scraping the motor on the stream bottom.

Except for portaging around unnavigable areas, late spring, summer, and fall travel has been exclusively by watercraft on open waterways, or on foot via land. In winter, when transportation is by dog team, snowmachine, and on foot, little bird hunting other than grouse and ptarmigan has occurred.

Blinds

Lime Village people have used both natural and handmade blinds extensively when hunting waterfowl. Blinds are called *k'uq'a*. Natural blinds include vegetation such as trees, brush, and grass, as well as hummocks, rock outcroppings, and other elevated landscapes. Hunters constructed spruce, willow, and other shrubs or brush, grass, and ice blinds, situating them in areas that waterfowl regularly use. Hunters use blinds for spotting birds and positioning themselves for a good shot but do not normally sleep in them. A blind may be used by several people at one time.

One type of constructed spruce blind consists of small spruce trees placed against one another tepee-style with a doorway facing open water used by waterfowl. Hunters also shoot through the top of the spruce blind and through tepee-style brush blinds. In the round brush-blind style, shots are fired through the top of the blind and through parted sides of the blind. Both spruce and brush blinds are made to look like standing plants and not, in one resident's words "like a city." Apparently no frame is made for either type of blind.

An ice blind is built on ice during the early part of the spring hunting season when the ice is firm and little open water exists. The ice blind consists of four sides and a flat, tightly laid pole roof. The three walls are formed by poles laid horizontally on one another. Grass is laid on the outer walls against which large chunks of ice are piled. An elder explains that the ice chunks must be stacked correctly as nothing else holds them in place. The fourth open side faces the direction where waterfowl are most likely to land. A small window is located in the ceiling and in one wall.

Grass blinds, located on lake beaches, are framed with straight, branchless stems of any kind of wood. Although fresh grass can be used, normally dry grass is piled against the sides and on the flat roof of the structure. Several long willow stems are tied over the grass to secure it. Willow is used because it is pliable and thus bends easily.

People have placed fish bones on the roofs of grass blinds to lure gulls. When the gull lands on the roof, the person positioned inside grabs the gull by the feet and kills it by twisting its neck. Gulls have been valued for their feather shafts. Grass blinds have been built at fish camps especially for this purpose.

Hunting Clothes

The Lime Village Dena'ina have used a variety of clothing when hunting birds. The choice of clothing depends on weather, travel, and other seasonal conditions. Camouflage, comfort, and the ability to move quietly are important qualities of hunting clothing.

Normally waterfowl arrive when snow and ice still cover a large area of the land and water. At this time hunters have traditionally worn white clothing in order to blend with the landscape, because birds are said to see colors that contrast with the landscape and are frightened by them.

White-haired Dall sheep skin parkas and pants have been worn to sneak towards birds such as geese, swans, and cranes found especially in wetlands. The hair on the clothes moves quietly on the snow and against brush or other objects. White snowshoe hare-skin (rabbit-skin) coats have also been worn for this purpose as have swan clothes whose feathers, like hare-skin fur, are quiet.

Lime Village people have used caribou parkas, pants, and boots with the hair on the outside for hunting waterbirds in spring snow and ice conditions. The skins are tanned in August when they are lighter in both color and weight and have shorter hair than in September. The whitish August caribou skins blend into the environment, and the lighter weight is more comfortable. Moose-skin clothes have also been worn at this time. In contemporary times, if skin boots were not available, hunters merely wrapped a piece of skin with hair outward around their boots so as to move quietly in the snow. White cloth parkas and white sheets have been worn for hunting birds in white environments.

A special kind of hunting leggings have been made from a single piece of moose leg skin. The top cut is made on the upper moose leg at the desired height of the legging, while the lower cut is made just far enough below the knee to fit the foot. The only seam required is one below the moose knee to create a fitted foot piece. The seam is waterproofed with grease and the hair is left on the outside of the tanned skin to allow the hunter to move quietly. Although used mostly for bear hunting, the leggings have also been worn when hunting waterfowl, especially in snow.

When the snow and ice no longer dominate the landscape, dark-colored clothing has traditionally been worn when bird-hunting to blend with the land and vegetation.

A number of kinds of waterproof clothing have been used for waterbird hunting including tanned, waterproofed caribou and moose-skin pants and hip boots when hunting waterbirds. The hair is usually removed before tanning the skin. This is waterproofed after the garment is made by rubbing grease or oil, often from waterfowl or moose, over the outer surface. Moose grease is applied especially to the seams. All waterproof clothing is sewn with a special stitch to also prevent water from entering the seams.

Other waterproof hunting clothing includes bear intestine coats, mittens, pants, and hip boots. Hip boots consisting of intestine pants sewn to short water-proofed caribou or moose boots are especially useful when wading to retrieve birds. A middle-aged person remembers his grandmother wearing such boots when he was a child.

Bear-gut clothing is said to be somewhat tougher than plastic; brown bear being stronger than black bear. The light clothing is easily carried and is worn just before entering the water. Wearing it through brush and other vegetation can cause it to tear. Waterproof fish-skin boots, pants, and coats serve a similar purpose and are said to have been used more frequently than bear intestine clothing.

Waterproof boots and other garments have not been sewn from featherless bird skin because the skin is too fragile and too stiff for comfort. Rain clothes have not normally been made from feathered-skin because the feathers do not stay waterproof long after they hit against something. However, waterfowl oil has been used to waterproof boots.

Methods of Calling Birds

Besides imitating bird calls to entice birds closer to them, hunters have used a variety of implements for the same purpose (see Learning about Birds). People use imitation and implements primarily for calling waterfowl. A resident observes that it is easier to imitate and lure swans and geese than ducks.

Traditional calling implements have been constructed primarily from plant and bird parts. Whistles for calling waterfowl have been made from dried cow parsnip *(Heracleum lanatum)* stalks (see P. R. Kari 1983, 93) and from spring willow stems when the sap runs and the bark peels easily. After selecting a piece of willow, one removes the bark with a knife or pounds the bark with a rock. The latter method loosens the bark so that it can be twisted from the wood. Pounding the bark also apparently softens the bark and prevents cracking during its removal. The deeper the wood is carved, the lower the sound produced by the whistle, while shallow carving gives a higher, squeakier sound. The mouth end of the whistle is flat and triangularly shaped. Whistles are called *viq' yidelyishi* in Dena'ina.

Blowing into the windpipe (trachea) with a sound box (syrinx) from harvested ducks has been another method of calling ducks. This organ is called *ts'en zitl'i* in Dena'ina. This type of caller can be used from a freshly killed duck (called *jija ts'en zitl'i)* or, if dried, can be soaked in water to make it work. Usually a duck will only respond to the caller of its species. When not in use, the caller is stored in water. Geese and swan windpipes (called *ndalvay ts'en zitl'i, tava ts'en zitl'i)* have served the same purpose as have callers made from certain bird bones (see Bone Technology). Empty shotgun shells *(vadruna)* have also been made into waterfowl callers.

Foods &
Products Made
from Birds

Preparing and Preserving
Birds for Food

Because waterfowl have been consumed in larger quantity than other kinds of birds, preparing and preserving them is described in the greatest detail. Although much of the same information pertains to other game birds, an attempt is made to note differences.

Lime Villagers consume almost all parts of a bird. People have traditionally eaten the meat, fat, bone marrow, organs, feet, the cleaned gizzard, and the head including the brain, eyes, and tongue. Due to lack of meat, the tip of the wing beyond the last joint is not used. One reason for not eating the bird intestines is that their small size makes cleaning them difficult. An additional reason given for not consuming waterbird intestines is that waterbirds may feed on poisonous water plants and that the poison could be transferred to humans. Bird feathers, bones, and beaks are not eaten.

Depending on the species, time of year, diet, and the species, bird bones differ in the amount of marrow or oil contained in them. For example, fall ducks are said to have more leg bone marrow than spring ducks since in the fall a person can suck the marrow from a cracked leg bone. A bird's fat content in other parts of its body also varies for the same reasons. People explain that birds have a large amount of fat in their shoulder and wing vicinity because of energy needed for flying. Waterfowl most favored for food have the highest general fat content including scoters, buffleheads, goldeneyes, and harlequin ducks.

Traditionally larger feathers have been plucked soon after the bird is harvested because the feathers are then most easily removed. Especially in earlier days, the birds were often needed for food immediately, a situation that has sometimes occurred in modern times. People give accounts of being at spring camps when freshly harvested birds were promptly cooked because there was little other food.

After waterfowl and sandhill cranes have been plucked, they are singed over a fire to remove the pin feathers and other remaining feathers. Skill is required to singe a bird so that the meat does not burn and acquire an unpleasant flavor. Singeing a bird is said not only to prevent waste of the skin, but in some people's opinion to improve its taste. Once the bird is singed, it is washed, butchered, and then cooked or saved for later use. At least one reason ptarmigan and grouse are not singed is because their feathers are more easily removed.

Boiling game birds for soup has continued to be the most common way of cooking them. Rice and potatoes are popular additions to the soup. For added flavor, birds may be smoked several days before being cooked.

Especially at hunting camps, people have frequently roasted gutted ducks and geese slowly on a stick over an open fire. Another traditional method of cooking waterfowl is to bake them in a hole lined with rocks dug in beach gravel. After wrapping the cleaned bird in birchbark, it is covered with gravel and a fire built on the gravel.

People have smoked gutted geese and ducks in the smokehouse for several days to a week both to flavor and to preserve them. If the weather is cool enough, the birds may be left hanging there or placed in a cache. They have also been preserved in airtight fish-skin sacks. When stored in a cache, birds taken in the early spring keep for about a month during cool weather and fall birds keep approximately until Christmas. The size of the birds and other factors may affect the length of storage. Another traditional method of storing birds so that they remain fresh tasting is by freezing them in water-filled birchbark containers.

Use of Bird Skin in Clothing and Other Products

The Lime Village Dena'ina have made two major types of cold-weather clothing from birds. One kind is constructed from feathered skin while the other style uses soft feathers as filling in skin and cloth. Birds employed for the former type of clothing include eagles, cranes, swans, geese, ducks, cormorants, and ravens. Swans, cranes, and eagles are said to have the strongest skins. Waterfowl down has most frequently been the filling in skin and cloth clothing.

Feathered-skin clothing such as parkas, pants, mittens, and hats, have been made primarily from the wingless bodies of the larger birds from which the tail and other large feathers have been removed. The skin is then tanned to remove the oil from the skin and to soften the skin.

Feathered-skin clothing has been lined with animal skin for additional warmth and strength. An elder explains that bird skin is fragile and tears easily without support. The furred skin of small animals has provided a warm lining for mittens and hats while caribou skins and the furred skin of a variety of animals have served as lining for larger bird garments. While the feathered side is the warmest side and worn against the body in the coldest weather, this reversible clothing is warm when worn either way.

People have created specialized clothing, such as jackets, particularly from the soft feathered necks and breasts of ducks but also from swans, loons, cormorants, and other waterbirds. This less hardy

TABLE 3. TYPES OF USEABLE BIRD SKINS

swan skin	*tava yes*
swan leg skin	*tava qa yes*
bald eagle skin	*ndatika'a yes*
bald eagle leg skin	*ndatika'a qa yes*
crane skin	*ndał yes*
crane leg skin	*ndał qa yes*
cormorant skin	*yeq yes*
cormorant leg skin	*yeq qa yes*
duck skin	*jija yes*
loon skin	*dujeni yes*
goshawk skin	*gizha kegh yes*
dipper skin	*tatsilqit'a yes*

clothing, especially admired for its attractiveness, is made entirely from one species of bird or from a variety of birds. If the breasts are used, they are always lined because breast skin is not as tough as neck skin, the most popular kind of skin used for this type of clothing.

An elder who wore feathered-skin clothing as a young person observes that when he came home from World War II, the practice of sewing bird skin clothing had been discontinued. He says that animal skin clothing continued to be used after bird skin clothing because the latter was harder to make. He reports that bird feather clothing was worn primarily as work clothing and not necessarily saved for special occasions. Although harder to produce than mammal skin clothing, bird skin clothing served as an attractive change from animal skin clothes as well as providing warmth and other utilitarian needs.

People indicate that due to the scarcity of large game animals in earlier times, feathered-skin clothing allowed the mammal skins to be used for essential items that feathered-skins could not provide.

Lime Village Dena'ina have continued in modern times to create feather-filled cloth clothing and bedding items. Although neck and breast down has been especially prized for it softness, medium-sized feathers have been used especially when down has been lacking or in short supply. Feather-filled articles have included snow pants, parkas, pillows, blankets, and other types of bedding and clothing.

Children's toys and art objects that closely resemble living birds have been fashioned from the feathered-skin bodies of waterfowl, woodpeckers, and other birds. With some exceptions, the bird re-

TABLE 4. TERMS FOR BIRDSKIN CLOTHING AND CONTAINERS

swan-skin coat	*tava dghe'a*
bald eagle-skin coat	*ndałika'a dghe'a*
swan-skin pants	*tava tl'useł*
bald eagle-skin pants	*ndałika'a tl'useł*
down feather blanket, robe	*k'keshch'a ch'da*
swan-skin robe	*tava ch'da*
swan-skin hat	*tava chik'ish*
swan-skin mittens	*tava gech'*
swan leg skin water container	*tava qa yes minłni viquhi*
crane leg skin water container	*ndał qa yes minłni viquhi*
cormorant leg skin water container	*yeq qa yes minłni viquhi*
dipper skin water container	*tatsilqit'a yes minłni viquhi*

mains intact including its head, feet, and wings. The bones are removed from the feet and the torso of the bird is removed through a cut on the side of one wing. The bird is stuffed with moss or grass because the plants absorb oil well (see Loons).

People have made sack-like water containers from the dried feet and lower, featherless leg skin of waterbirds such as swans, cormorants, dippers, and sandhill cranes. Swan skin followed by crane skin produce the biggest and thus the best containers. The sack-like containers are sewn with a special waterproof stitch and are tied on top with leather or other available string. The containers have primarily been used when traveling.

Feather Technology

The Dena'ina classify several types of feathers. There are two general terms. Long feathers with a shaft (contour feathers) are called by the stem word -t'u. Down feathers are called by the stem word -keshch'a. The soft under feathers, down feathers and the tiny hair-like pin feathers (filoplumes) are dituxi. Plucked down, as well as fine fur is called k'kidza. The stiff wing feathers (flight feathers) are k'ch'enlna t'u and tail feathers are k'ka t'u. Some people recognize the head feathers and crowns of feathers as a separate category (vetsiduq' k'andazdlits'i).

The following is an account of the upper Stony River Dena'ina feather technology. (See also the feather clothing described in Bird Clothing.) Flight feather shafts of eagles, swans, cranes, cormorants, gulls, and other birds with strong shafts have been used to make snares, fishing equipment, and other items. Although eagle feathers are said to be the strongest, gull feathers appear to have been most commonly employed because they are both strong and easily accessible. Besides their strength, feather shafts are valued for being flexible, lightweight, and waterproof. An elder describes them as having the qualities of plastic. They can be used either fresh or dried.

Lime Village people have made technological items from fresh or dried feather shafts by removing the barbs that make up the vane of the feather and splitting the shaft lengthwise to the desired width. To split a shaft, a small cut is made in one end of the shaft. The shaft is separated by holding one part of the shaft by the teeth and pulling slowly down by hand on the other section. The marrow is then removed from the shaft.

Bird, fish, ground squirrel, and other snares for small animals have been made from feather shafts. Due to their exceptional strength, golden eagle flight-feather shafts have been used to snare lynx. Using a special knot, dipnets, set nets, "gunnysack" nets, and scoops for harvesting fish have been woven from feather shafts. Gunnysack nets have served as part of both beaver and fish traps. For all items, elders emphasize the importance of making the correct knot well. The distance between knots determines the mesh size and thus the type of prey caught in the net.

Although weaving nets and scoops with feather shafts is considerably more time consuming than making them from other materials (such as sinew and twisted willow bark), their long life and light weight is said to be worth the extra time. Ideally nets and other items made from multiple feather shafts consist of shafts from one species. However, if enough of one species is not available, the item can be made from feather shafts from several species.

Miscellaneous uses for feather shafts have included lashing for sewing birchbark canoes (though spruce roots have been preferred). Birchbark baskets have been decorated by weaving them with dyed feather shafts. Red and black ochre provide the best dyes for feathers because they are said to be more durable than berry dyes.

Straight wing or tail feathers of large birds have been attached to arrow shafts so that the arrows fly straight. If necessary, the feathers are cut to the correct size and shape. Because they are soft, owl feathers are only used when other large bird feathers are not available. The water resistant feathers of waterbirds, including those of cormorants, have been used on arrows for hunting waterbirds and, during rain, land birds. Large land bird feathers such as eagle feathers have been employed for hunting land birds in dry conditions. The feathers of gray jays and magpies, for example, are too small. Both the feathers and arrowhead points have been tied with sinew to the arrow shaft. Whole wings of ducks, geese, and other large birds have been used as brooms and sweeping tools *(veł qelchezhi).*

Eagle feathers have been worn on the headdresses of singers at potlatches and possibly by warriors. Shamans have worn feather headdresses (J. Kari 1977, 239). Brown bear skin headbands contained bald eagle, golden eagle feathers, or a combination of the two kinds of feathers. Although golden eagle feathers are harder to obtain than bald eagle feathers because golden eagles live in the mountains, the two types are said to have equal value.

Bird Bone Technology

Bird bones have served the Lime Village people in a variety of ways. Both men and women have made beads from duck wing and leg bones and from geese and swan foot bones. After they have been boiled and cleaned, the warm bones are cut to bead size. If the bones are cut when cold, they may chip. In early times, a hard, sharp, shiny brown rock was used to cut the bones.

The beads have been colored with various plant dyes: alder bark—red; rotten willow wood—blue; rotten spruce wood—reddish brown; and berries—a variety of colors. The beads, soaked or simmered in the dye, have been used in necklaces and earrings and have been sewed on headbands, coats, and other clothing. When a bead is sewed on a fur parka, the fur is partially cut from the area where the bead is to be placed so that the bead is visible. See Swans for information on windpipe beads.

Lime Village people have made whistles, *viq' yidelyishi*, for calling birds and animals from cooked, cleaned bird bones. An elder reports that his father made many different kinds of whistles. The sound of the whistle varies with the kind of bird and bone used, the size of the bone, and the number of holes made along the length of the bone. Depending on the desired sound, the caller blows into one end of the bone while covering the other end of the bone and perhaps some of the holes along the length of the bone.

Swan wing and leg bones have been used as "straws" *(ts'en zitl'i)* and have been left at locations regularly used for obtaining drinking water. Split wing bones of various birds have served as toothpicks *(ghitl'en ts'detseyi)*.

Medicinal Uses of Birds

Few Lime Village medicinal uses of birds have been documented. Fresh raw or rotten bird meat and other animal meat has been secured on an infected or blood-poisoned area to draw the sickness from the area into the meat. The meat is left on the area until it is very rotten because rotten meat is said to heal the area faster than fresh meat. Any kind of soup including duck, grouse, and ptarmigan can be given to a sick person to increase his appetite and provide nourishment.

Taming & Training Birds

The upper Stony River people's tradition of taming birds has met a variety of needs. Not only do tamed birds give enjoyment and companionship, certain species also provide protection and hunting assistance. The processes of taming, training, and caring for birds also allows a special opportunity to observe and learn about them. Observing birds closely in these ways was undoubtedly especially important for identification purposes before binoculars were available.

Birds have been tamed in a number of ways. One common method mostly used in the spring and summer seasons is to simply place food near the house or camp site to coax the birds there. Each day the food is moved closer to the living quarters and often eventually inside as the birds become more relaxed around humans. Smokehouses and other out buildings are also used for this purpose. The birds, frequently small or medium-sized land birds, are generally allowed to come and go at will. Sometimes they assist people by eating flies and other noxious insects. Birds tamed this way include robins and other thrushes and sparrows.

Traditionally children especially have live-trapped small birds with snowshoes, birchbark baskets, and more recently wash tubs and other modern containers. Nets have sometimes been placed over the traps to insure that the bird does not escape. A snowshoe trap is made by placing one snowshoe flat on the ground and holding the other snowshoe up on one end by a stick at an angle above the other snowshoe. Gloves are placed on both snowshoes to pad them and bait is put on the bottom snowshoe. When a bird enters the trap to feed, a person lowers the upper snowshoe onto the bird by pulling a string tied to the stick.

Birchbark baskets and other containers used as traps work in a like manner except no padding is necessary. Traps have often been placed in a smokehouse where birds tend to feed on insects attracted by drying fish. Cooked maggots are a traditional bait. Children watch the birds for a time before freeing them.

A fall tradition has been to catch small migratory birds that do not leave with the other birds. People have then taken them into their homes and cared for them as pets during the winter. They have enjoyed the birds both for observation and for companionship. The birds are said to sing very softly in the winter and usually before guests arrive or during their visit. When warm weather returns, the birds are allowed to come and go as they desire. Normally they migrate with the other birds the following year.

People have also kept stranded migratory birds alive during the winter by feeding them outside and note that they become especially large and fluffy. One person's explanation for migratory birds remaining behind is that they overeat and become lazy.

Young birds that are able to eat by themselves have been removed from nests and raised as pets. An elder remembers that when he was about ten years old he took two ravens from their nest and raised them. He made a pole fence constructed like a fish rack for them to perch on and placed grass on the ground underneath where they would fly and rest. He fed them meat and crackers when he first took them and found that keeping them fed was a difficult task. He talked to them as a person talks to a dog and says they quickly learned to understand him. Soon they were following him wherever he went and became his hunting assistants.

Traditionally ravens, hawks, and other birds of prey have actively been trained when young to sight and hunt game for people and alert them of danger. A bird circles around an area where game is located to alert his trainer. It flies back to the person and calls in a certain manner for the same reasons. Some birds will capture game for the person. They learn like people, that is by watching what happens and listening to what they are told. Untrained tame birds that accompany a hunter and wild birds may unintentionally alert hunters of game or danger but are generally less efficient than trained birds.

People frequently mention that one-week-old geese have been taken for pets when found swimming in streams and lakes, one or more goslings being caught at a time for a single person or family. The bird(s) would quickly attach itself to the family or person that raised it. White-fronted geese have been preferred because they are said to be smarter than Canada geese that have also been tamed. Some ducks such as mallards are good pets while certain birds including grouse die in captivity.

A tamed goose might migrate south with other geese in the fall, or it might stay with a family throughout the winter. One person who recalls having a Canada goose as a family pet says that it slept in the house in a box during the winter and outside in the summer. It would not take food from the house unless given it. The bird would not fly far by itself and would accompany her and members of her family when they went somewhere. They would talk to it like a dog, and it would understand them. Among other food, winter "house" geese have been fed crackers, rice, and fish.

Adults remember that when they were children their pet geese would follow them in their play and work. They were not only pets but also protectors, or as one person called them "official guardians." For example, geese watching for danger would circle over and land next to children who were picking berries. Vonga Bobby's account of how tame white-fronted geese saved women and children from a brown bear is on page 39.

Geese have been trained to spot game for hunters. An elder explains that a trained goose can find game for its owner or other familiar people if told to do so.

A person who raised hawks and owls as pets took the young birds from the nest before they could fly. The parents would protest but not attack him when he climbed the tree and dropped the chick to the ground. He observed that they continued to care for the other young but would move the nest the following year. He tethered the captured bird until it got used to him, and then he allowed it to fly at will. The bird ate food given it and hunted for itself. In the fall, it departed with the other birds.

It is said to be bad luck if a bird accidently and unwillingly enters a house. It can also be unlucky to touch a live wild bird that has not been tamed. Exceptions include live-trapping birds to observe them, taking young birds to tame, and assisting or killing injured birds.

How Tame Geese Saved the Lives of a Woman and Her Two Daughters

by Vonga Bobby

Long ago people raised wild geese at Htsit (Tishimna Lake area). There was no large game in the area at that time because of a forest fire. A mother and her two girls went away from the village to pick berries. They started picking berries, then made a fire to have tea. They split whitefish and put the pieces by the fire to roast. The fish oil dripped into the fire and the smoke went towards the Stony River. The two tamed white-fronted geese with the woman and her daughters went a distance from them and hollered to alert them.

There was a brown bear down there somewhere that scented the burning grease and the whitefish. The bear went toward the smell and came out fairly near the women. The bear started toward the woman and girls and because they did not have any way to defend themselves, they did not know what to do. The bear kept coming towards them and they became frightened. When the bear was quite close to them, the geese took off and started circling the bear. Then the geese landed near the bear. As soon as the bear raised up and tried to catch the geese, they landed a little further from the bear. While the geese were doing this, the daughters and mother went over the ridge where the bear could not see them and then went to the village.

Soon after, the mother and some other people went out and heard the geese really hollering. The geese ran and met the people and encircled them. Then the geese went back to near the brown bear. This was the second time the geese gave assistance. Someone said, "I should kill the bear," and they ran toward it. The bear was very smart and ran away. They watched it for a long time while it ran away.

That's why they liked to raise geese because the geese helped them. The geese saved the woman and her daughters' lives.

(This is a figurative translation by Priscilla Russell from the transcribed text with literal translation by Luthur Hobson Sr. and James Kari.)

Beliefs About Birds

Traditional stories describe an earlier time when all living beings were different kinds of people and could speak the same language. Among other kinds of people, there were the bird people, the insect people, the animal people, and the fish people. For example, chickadees, loons, owls, and other birds known today were bird people. Over time events occurred that changed the different kinds of people into the forms known today. Since that distant time, human people have respected the various forms of life and the environments that sustain them all.

The Dena'ina show respect to the natural world and its life by their actions. Preventing waste of a resource is a very important tradition. They do this by not harvesting more of a resource than is needed and by using as much of the resource as possible.

When harvesting a bird, hunters make an effort to kill it instantly to prevent suffering. They make an extended attempt to regain a wounded bird both to avoid suffering and waste. The Dena'ina custom of sharing food prevents waste and allows people to enjoy freshly harvested food more frequently than if they did not share.

A young adult hunter states that he cannot imagine a world without birds, especially the ones he hunts. Because of this, the decision whether to shoot or not shoot is each time an extremely important one for him. His value system reflects not only that of his own generation but of countless past generations.

According to Dena'ina belief, a bird allows a person to kill it. The bird eventually lives again as the same kind of bird that it was before its death. This process is said to be true of animals and other kinds of life.

People involved in the death or use of the bird are responsible for correctly disposing of the bird's remains. For example, the bones of waterbirds should be returned to water and the bones of land birds left on land under a tree or in another secluded area. The bones should never be left on a trail. Taking correct care of the bones shows respect for the bird and aids in its quick return to life. If the bones are disposed of carelessly, the bird has a slow, difficult time coming back to life. Not only does this cause needless suffering for the bird, it means fewer birds and less food for people.

If a bird is found dead, it should be treated in the same respectful way as described above. For example, a bird killed by hitting a window should be given proper care. Lime Villagers still practice these customs.

Communication with Birds

The Dena'ina believe that birds use a wide variety of vocal and non-vocal actions to communicate among themselves, with other wildlife, and with humans. Bird expression has held great significance for the Dena'ina who have traditionally communicated with birds. This discussion emphasizes vocal communication.

Lime Villagers especially enjoy the mating (or territorial) songs of the migratory birds when they return in the spring and early summer. The songs are a refreshing addition to the sounds of permanent bird residents. Elders explain that the birds sing at this time because they are happy to be back in Alaska, the place of their birth. One hill northeast of Lime Village called Shan K'denshisha is said to mean "summer sounds" and refers to bird songs. According to at least some elders, the birds find mates by sight, not by sound. Once they are nesting, they make alarm sounds or "holler" if danger is near their nest. Members of the same species and different species may verbally warn each other in nesting and other situations. For example, geese in the same vicinity make alarm calls to one another that other birds understand. The yellowlegs, known locally as "the tattle tale bird," especially irritates hunters because of its loud calls when it spots intruders. As has been traditionally true of human beings, birds are said to have medicine songs for hunting and other purposes.

Birds heard by people in certain specific circumstances have special significance for the people. The type of sound may but not necessarily be essential to the situation. For example, if a loon calls near the village, it may mean that misfortune will occur. When the olive-sided flycatcher calls *vava* in the early summer, it foretells the arrival of salmon. In the Dena'ina language, *vava* means "dried fish." Some birds are said to be able to communicate with a person in the person's language.

In conversation, elders especially may refer to a bird by imitating its sound whether or not the bird is actually named after one of its sounds. People are expert at identifying birds by their calls and songs. Lime Villagers, often as children, have learned to imitate bird sounds for calling birds closer when hunting them. They learn both by listening to the bird sounds and hearing hunters use them.

The Dena'ina language has a rich vocabulary for describing sounds. The sounds of birds are reflected in the names for birds in several ways.

A. Some bird names use a special onomatopoetic verb describing its call:

trumpeter swan	*dult'iya*	'the one that calls *t'iy*'
spotted sandpiper	*delvizha*	'the one that calls *vizh*'

B. A few names are compounds that describe a bird and its sound

Wilson's snipe	*yuzíł*	'sky whistler'
savannah sparrow	*ch'ich' qunsha*	'ground squirrel that goes *ch'ich'* (scraping noise)'

C. Some bird names are *ideophones*, words that imitate the call of the bird and are the name of the bird at the same time:

long-tailed duck	*ahhanya*
American wigeon	*sheshinya*
gray-cheeked thrush	*ggezhaq*
golden-crowned sparrow	*tsik'ezdlagh*

In addition, there are few special verbs for raven calls: *delggagh* and *delgguq*. Also there are many phrases that characterize bird songs (and our documentation of this is not very extensive). For example, one call of the raven is *ggagga* (animal). A call of the golden-crowned sparrow is *chik'a dulnił* (there will be firewood).

Birds also communicate with people by signs. For example, a strange action of a bird may predict a negative occurrence. On the other hand, predatory and scavenger birds circulating above may indicate game in the area.

PHOTO GALLERY

Some caribou grazing near Vił Qutnu Dghil'u "caribou snare mountain," Cairn Mountain, about ten miles south of Lime Village. Photo by James Kari.

Lime Village viewed from the air at breakup in May of 1992. Photo by Priscilla N. Russell.

Nizdlu Dghil'u "islands are their-mountain," in the northerly group of Lime Hills, is a upland area about seven or eight miles north of Lime Villlage. Photo by Priscilla N. Russell.

This roundish rock on the Stony River is on the edge of the southerly group of the Lime Hills. The name Q'in Tetl'i "exploded fish egg" is one of the most charming and distinctive Dena'ina place names. Photo by Priscilla N. Russell.

A slough of the Stony River below the village with Q'in Tetl'i "exploded fish egg" in the distance. Photo by Priscilla N. Russell.

Tundra Lake, locally called "606," is Vendash Vena "shoal lake-lake." This has been a popular area for spring and summer waterfowl hunting. Photo by Priscilla N. Russell.

Htsit Vena "lowland lake" Tishimna Lake looking to the east with the Lime Hills in the distance. Photo by Priscilla N. Russell.

Luther Hobson with a view toward the north from Chida Dghil'u "grandmother's mountain." Photo by Priscilla N. Russell.

Evan "Bimbo" Bobby viewing the terrain near Htsit Vena "lowland lake" Tishimna Lake. Photo by Priscilla N. Russell.

Pete "Fitka" Bobby standing near the old village of Qeghnilen "current flows through" on the upper Stony River. In the distance is the mountain called Chaqenq'a Qeghneq "upland of the smokehouse." Photo by Priscilla N. Russell.

Katherine Bobby and Joe Bobby studying the landscape near Hek'dichen Hungry Creek. Photo by Priscilla N. Russell.

Nora Alexie in August of 1994. Photo by James Kari.

Pete Bobby watching the channel of the Stony River. Photo by James Kari.

Emma Alexie shown here dipnetting for whitefish on Hek'dichen Hungry Creek in September of 1982. Photo by Priscilla N. Russell.

Priscilla Russell and Emma Alexie on the side of Q'in Tetl'i "exploded fish egg" in 1981. Photo by Priscilla N. Russell.

Nora Alexie with a rack of furs. Photo by Priscilla N. Russell.

Nick Alexie heading north from Lime Village with his dog team. Nizdlu Dghil'u "islands are their-mountain," the northerly Lime Hills are in the distance. Photo by Priscilla N. Russell.

CLASSIFICATION OF BIRDS

Classification of Birds

D ue to the complexity of folk classification systems, this section only partially documents how the upper Stony River people classify birds. The divisions in this section are an attempt to represent at least a part of the Dena'ina method of bird classification. Bird categories are based on shared physical, behavioral, environmental, or seasonal characteristics. For example, phalarope is thought to be grouped with the two species of grebe. A bird's role in traditional stories may also influence how it is grouped with other birds. As is true of other folk systems, a bird may fit into more than one category on the same level and thus be associated with members of several categories.

The Lime Village Dena'ina appear to consider birds a unit of the animal world as opposed, for example, to the insect, fish, and plant kingdoms. The stem word *ggagga* abstractly seems to mean "creature," and it occurs in several mammal names. *Ggagga* is both a general name for four-legged creatures and a specific name for brown bear. In summer the red fox is called *ggaggashla* "little creature." *Ch'ggaggasla*, the general name for bird, may mean "someone's little creature." The chickadee is called *ch'ggagga*. Note that the word *ggagga* is not used for other types of beings such as fish or insects. Both mammals and birds are warm-blooded creatures, and it appears that mammals and birds are closely associated in the Dena'ina classification system.

Birds are differentiated from mammals, fish, and insects primarily by their beaks, feathers, wings, two legs, and other unique body characteristics. See bird anatomical terms in the first section.

When asked if a *heł jech* "little brown bat" *(Myotis lucifugus)* was a kind of bird, elders responded that bats have fur and teeth and are definitely not birds.

The broadest division among birds is between permanent residents *(hey ch'ggaggashla* "winter birds') and migratory birds *(shan ch'ggaggashla* "summer birds'). Two kinds of migratory birds are noted—those that remain throughout the summer and those that pass through the area, usually in the spring or fall (see Seasonal Cycle).

Apparently all of the small winter birds *(hey ch'ggaggashla)* are considered to be relatives including those that remain only part of the winter. Winter birds seem to be those that reside in the area at least most of the time when snow is on the ground or are only in the area when snow exists. However, at least some people appear to recognize a category of spring birds, those that arrive when snow exists but remain throughout the summer. Regardless, a true winter bird appears to be one that resides in the area throughout most or the entire winter. Most nomadic birds qualify as winter birds.

The Dena'ina sometimes call the most important member of a category by the same name as the category to which it belongs. For example, *jija* is both the Inland Dena'ina general name for "waterfowl" and for the class of waterfowl "duck." Similarly, *q'ach'ema* is their general name for "ptarmigan" and more specific name for "willow ptarmigan."

The Lime Village Dena'ina recognize a number of major bird classes determined primarily by physical traits or environmental needs. When known, Dena'ina class names are included. The classes appear to be (1) Waterbirds; (2) Ospreys, Eagles, Hawks, Hawk-like Birds, and Owls; (3) Grouse and Ptarmigan; (4) Cranes and Shorebirds; (5) Gulls and Gull-like Birds; (6) Climbing Birds; (7) Scavengers; (8) Small Birds (sub-categories consist of winter birds and summer birds; and (9) Nomadic (Wandering) Birds that the authors have introduced as a category.

Nomadics—wandering birds—are species that do not come and go on the same schedule each year. In one year they will come in large flocks, may remain to nest, and move on. The next year, this species will be absent altogether. The occurrence and breeding of nomadic species depends largely on the seed crop of trees (spruce, birch) and shrubs (alder). Within each of these classes we have arranged the species according to the taxonomic system used by Western scientists (AOU 1998).

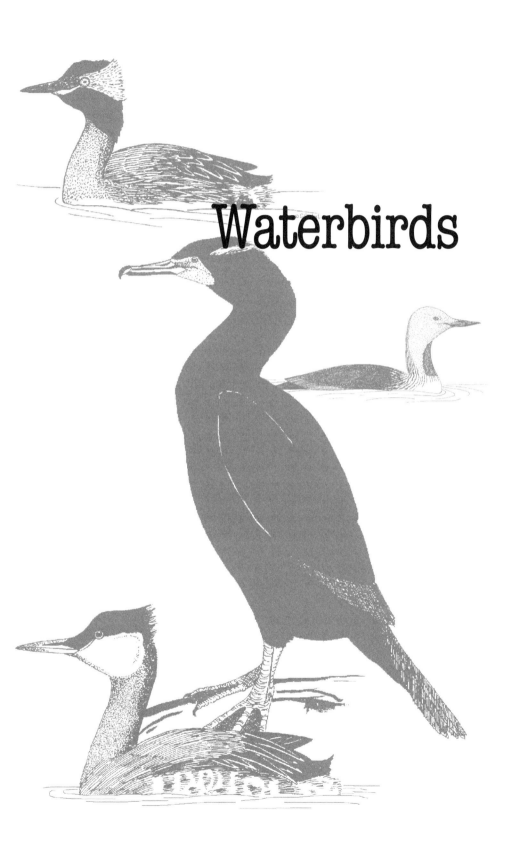

Waterbirds

Loons: *Dujeni*

Loons are large diving birds with a long sharp bill. They catch fish by swimming underwater using their large webbed feet. Because their legs are far back on the body, they do not walk easily on land, but waddle or slide to and from their nest which is usually situated near the water's edge. When flying, loons often hold their head and neck below the horizontal; their feet project beyond the tail in flight.

RED-THROATED LOON *(GAVIA STELLATA)*

shdutvuyi possibly "gray bill"

Description: Sexes alike. Smaller than the common loon, has thinner and upturned bill, often holds head with bill tilted up.

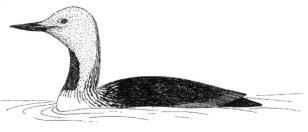

Summer: head and sides of neck pearl gray, nape striped, throat dark red, back dark and unmarked. Winter: gray cap, white face including around the eye, back speckled, pale below.

Pacific Loon *(Gavia pacifica)*
(formerly classified with Arctic Loon)

ggulchun "helpless runner"
also called *ggagga dik* "creature ?"

Description: Scientists recently divided the arctic loon into two species: the arctic loon of the far north Bering Sea, Arctic Ocean, northern Alaska, Canada, and Eurasia and the Pacific loon of western North America. Sexes alike. Smaller than common loon, has thinner straight bill, holds head horizontally. Summer: head pearl gray, throat dark (iridescent purple in good light), fine streaks on sides of neck, back checkered black and white. Winter: dark cap, white face below the eye, back solid dark, white below, often has indication of neck ring at throat.

Common Loon *(Gavia immer)*

dujeni "?" the name may refer to the loon's call

Description: Sexes alike. Large loon with heavy black bill. Summer: dark head and neck with black and white striped collar or necklace, checkered black and white back, distinctive yodeling call. Winter: dark cap, white face usually including eye, dark unmarked back, heavy bill is paler in winter, often gray. A traditional Dena'ina story explains how the common loon receives its dentalium necklace. A version by Peter Kalifornsky appears in Kalifornsky 1991, 144–149.

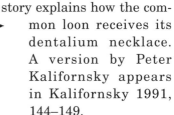

WATERBIRDS

During the open water season, the red-throated, Pacific, and common loons inhabit lakes and rivers of the upper Stony River area. Loons return later than other waterbirds after the lakes and creeks are largely free of ice. Loons do not sit on the ice as do many other waterbirds.

Lime Villagers admire loons for their diving ability and say that the Pacific loon is an especially fast diver. They respect loons for their strong bills that are strong enough to puncture the underside of skin, birchbark, and canvas canoes and to make large dents in wooden boats. In earlier days, people have drowned due to loons damaging their canoes. The birds are especially aggressive in the vicinity of their nest.

Although loons have not normally provided food for Lime Villagers, their feathered skin has supplied clothing and bedding covers.

Dried loon skins consisting of the body, head, feet, and wings were stuffed for use as decorative house objects and as children's toys. To serve as a container, the body of a decorative bird may be only partially filled and may have a lid. A Lime Village resident remembers that a stuffed common loon kept in the house was enjoyed for its beauty and that the white throat spots, which remind people of a dentalium necklace, were especially pretty. People recall that a toy common loon looked much like a real loon. They say that the traditional belief that the owner of a toy would grow up preferring to travel in boats rather than on foot in the mountains has proven true. Toy birds have also been crafted from wood.

If a loon, especially a common loon, circles around a village and "hollers," it is said to warn of coming misfortune such as a person's death. However, a Pacific loon calling often along a river predicts abundant fish while a common loon doing the same thing may be a bad omen. The reason is that Pacific loons "like" rivers as well as lakes but a common loon's home is only in lake country. Another traditional belief is that windy weather will occur if a loon makes a howling sound similar to that of a wolf. An elder comments that common loons call under water in the same way that they do in the air and that loons fly under water.

Grebes

Grebes are diving birds with lobed, rather than webbed feet. They swim underwater to catch fish and insects. Like loons, they are awkward on land and build their nest often on a floating island of vegetation. They spend summers on fresh water lakes and ponds and winter usually in the ocean or larger lakes that do not freeze.

HORNED GREBE *(PODICEPS AURITUS)*, "MUD DUCK"

nachandghelahi "the one that catches our scent"

Description: Sexes alike. A small grebe with a sharp pointed grayish, not yellow, bill. Summer: black head with conspicuous buffy ear tufts, dark red neck and sides, dark back. Winter: dark cap, white face and throat, dark back.

The Dena'ina name describes their impression of the horned grebe's sense of smell.

RED-NECKED GREBE *(PODICEPS GRISEGENA)*, "MUD DUCK"

taqa'a "water foot"

Description: Sexes alike. Distinguished from the much smaller Horned Grebe by the long yellow bill at all times of year and by its white cheeks in summer. Summer: white

cheeks, dark cap, and rusty red neck, dark back. Winter: dark above and light below, best identified by bill, shape, and habits.

A possible reason for *taqa'a* meaning "water foot" is the grebe's strong swimming ability and awkwardness on land. Their local English name "mud duck" refers to their preference for swampy lake areas especially when nesting.

The Lime Village Dena'ina have not normally harvested grebes for food but have used their skin and feathers as described for other waterbirds. Grebes are said to be closely related to phalaropes.

Phalaropes

RED-NECKED PHALAROPE *(PHALAROPUS LOBATUS)*, "MUDSUCKER"

taqa'a vekela "water foot's younger brother"

Description: Phalaropes are small shorebirds that spend most of their time on the water's surface where they often swim in circles attempting to stir up underlying sediments to bring insect food to the surface. Female phalaropes have brighter plumage than males because the male incubates the eggs and raises the young.

Adults have a gray to brown head, white throat, dark red stripe down the neck, brown back, and white belly. The bill is needle shaped. In flight, there is a conspicuous white wing stripe. The toes are lobed and not webbed.

The red-necked phalarope is described as chickadee-size and as closely resembling the grebes, the reason for their being related to the grebes in the Dena'ina classification system. Phalaropes are said to arrive in April with the waterfowl.

While their local English name "mudsucker" refers to their obtaining food from wetlands, they are known for their unusual swimming ability as described above. Because two Dena'ina names exist for the red-necked phalarope, it is possible that the sexes are given individual names. Although unlikely, it is possible that a second phalarope species occurs in the area (red phalarope, *Phalaropus fulicaria*). No use has been recorded for phalaropes except like almost any bird, they can serve as emergency food.

Cormorants

Cormorants are large dark diving birds that swim underwater using their large webbed feet and large bill with expansible throat pouch to catch fish. They often perch on shore or in trees with their wings held open as they dry their plumage. Cormorants often swim with only the head and neck above the water. The Dena'ina apparently do not classify cormorants with *jija* (waterfowl).

DOUBLE-CRESTED CORMORANT *(PHALACROCORAX AURITUS)*, "SHAG," "BLACK SHAG"

yeq, yeq cheh √ and
"big cormorant"

Description: Sexes alike. Birds take two years to reach mature plumage. Adult: throat pouch is yellow or yellow orange; in summer, breeding birds have white tufts on sides of the head. Juvenile and first year birds are dark brown above and lighter brown below; the face and throat are dull yellow.

Two names have been recorded, *yeq* and *yeq cheh*. Formerly we thought these may be two species, one being pelagic cormorant *(Phalacrocorax pelagicus)* that is smaller than the double-crested cormorant. The pelagic has a dark red face and dark bill unlike the yellow face of the double-crested. In breeding plumage, it has conspicuous white flank patches and white plumes on the neck. It is common in coastal Alaska and rarely strays from salt water. Probably "big cormorant" and "cormorant" refer to the mature and immature double-crested cormorant.

At one time a colony of three to four hundred double-crested cormorants nested regularly above Lime Village in a canyon area along the upper Stony River. The canyon has the place name Yeqtnu Denyiq' "cormorant stream canyon." A group of five place names in this area refer to cormorants. People report that the large number of nests situated on rocks in a swift area of the river were eventually washed away by high water. They observe that the birds moved further downriver but are occasionally seen flying near their old nesting site.

Lime Village Dena'ina have made feathered skin clothing and bedding from cormorants. An elder specifically mentions very warm, feathered-skin cormorant parkas. Wearing a cormorant feathered-skin hat as a young person to prevent early gray hair has been a custom. People have sewn waterproof containers from cormorant feet and leg skin as they have from swans and other large waterbirds.

Elders remember eating cormorant that they say had a "fishy" and not especially pleasant taste. Cormorants have not normally been hunted for food, although their eggs, which are said to taste like duck eggs, were regularly harvested when the birds were abundant in the area.

A traditional belief is that a cormorant flying and calling around a village or house warns of possible misfortune.

Waterfowl

Swans: *Tava*

Swans are large, long-necked, all-white waterfowl that swim using large webbed feet. They tip up and search for underwater vegetation using their long necks. Swans migrate in large flocks. Tundra swans migrate through Interior Alaska but nest on the northwest coast. Trumpeter swans nest throughout Interior Alaska.

TUNDRA SWAN (*CYGNUS COLUMBIANUS*)

tava "water-white"

Description: Sexes alike. Base of bill has a small yellow spot. The neck is often carried in a curve, and there is rarely any rusty discoloration to the head and neck.

TRUMPETER SWAN (*CYGNUS BUCCINATOR*)

dult'iya "the one that calls *t'iy*"

Description: Sexes alike. The bill is all black. The neck is usually held straight up without a curve. The head and neck are often rusty colored from iron oxide in the water in which they feed. Trumpeter swans have a loud trumpeting call which gave rise to the English and Dena'ina names.

Swans, the first spring waterfowl arrivals, usually reach the upper Stony River area by mid-April. The trumpeter swan and possibly the tundra swan breed in the lake country near Lime Village. A swan, species not known, has been seen in the Lime Village area in January.

During the spring and fall, people have harvested swans with snares and guns. Great effort has been made to shoot the bird's head because hitting the wing only breaks it and causes the bird to suffer. Elders say that bow and arrows were not used to kill swans for this reason.

Lime Villagers have expressed a particular appreciation for swans and appear especially sensitive concerning harvesting them. People have repeatedly expressed appreciation for their beauty and that they mate for life. They prefer to take only a few during the hunting season because they enjoy the taste. The birds are often shared at potlatches and other festive occasions.

However, swans have been an important food source in the spring when food was scarce. Not only do swans return earlier than other waterfowl but because of their larger size, one swan provides significantly more food than a goose or a duck. As is true of other waterfowl, Lime Villagers have traditionally consumed almost all parts of a swan including the outer part of the gizzard after it is cleaned of gravel. They have gathered swan eggs for food only when they have lacked other food. Customarily, swans and their eggs are normally boiled for consumption; they are not preserved for food as is the meat of other waterfowl.

Feathered swan skin has provided clothing and blankets. Swans, eagles, and cranes have the sturdiest skin of all birds used for clothing. Jackets and other clothing admired for their beauty have been sewn from swan necks. A Lime Villager remembers her deceased grandfather wearing a feathered swan skin hat made from the head and neck. Lengths of black skin from swan legs were used as striping in clothing seams. Swan down, said to be the warmest down, has been used for bedding filling and cold weather clothing filling. The best bird leg and foot skin water containers have been crafted from swans.

Hunters have employed swan windpipe bones *(ts'en zitl'i)* for calling birds closer by blowing into the instrument. An elder comments that the caller works well. Swan windpipes have been worn by boys so that they would develop good breathing capacity. Beads have been made from swan windpipes.

Swan wing and leg bones cut with holes on both ends have served as drinking straws. They have been placed by streams and lakes for sucking water. Girls menstruating for the first time sipped water from a swan bone tube.

Lime Villagers have employed swan grease to waterproof skin footgear and skin canoes and kayaks.

Geese: *Ndalvay*

Geese are large waterfowl with webbed feet. They tip up to feed on underwater vegetation using their long necks and graze on grasses and other emergent vegetation. People have observed that the goose population has declined since early days.

GREATER WHITE-FRONTED GOOSE *(ANSER ALBIFRONS)*, "SPECKLED-BELLY GOOSE"

k'dut'aq'a possibly "bib" or "chest"

Description: Sexes alike. Generally appearing as a gray goose when compared with the snow and Canada goose, white-fronts have a pink bill, white patch at the base of the bill, and black markings on the belly,  giving rise to the colloquial name "speckle-belly" and to the Dena'ina name.

Snow Goose (*Chen caerulescens*)

ch'iluzhena "black wing"

Description: Sexes alike. Smaller than the other geese, the snow goose in Alaska is all white with black wing tips (Dena'ina name). The bill is dark pink.

Brant (*Branta bernicla*)

chulyin viy'a "raven's son"

Description: Like a small Canada goose, the brant is brown above and light below with a dark chest, head, and neck. There is a series of fine white markings on the neck. Normally restricted to coastal areas, brant sometimes move up major rivers and have been seen on the Stony River.

Canada Goose (*Branta canadensis*)

ndalvay "the one that is gray-spotted"
also called *ventl'u ch'anlch'eli* "the one with light-colored cheeks"

Description: Sexes are alike. Recognized by the black head and neck with distinct white chin patch (Dena'ina name). The back is brown.

Geese are spring migrants to Lime Village lake country, normally arriving after the swans but before the ducks. The majority of greater white-fronted and snow geese pass through the Lime Village area after feeding there. Remaining geese nest in the wet lowlands near the lakes and on the lake shores. After nesting, geese are commonly found on rivers and smaller streams with their goslings.

Lime Village people have snared geese and shot them with bows and arrows and more recently with .22 rifles and shotguns. Hunters note that geese are smarter than most other waterfowl in knowing how to stay out of shooting range. They have obtained them best by skillfully hiding and sneaking towards them and by calling to them.

Geese and their eggs are relished for food as are most geese body parts. They have made goose feathered-skin clothes and used their down feathers for clothing and bedding filling.

Besides the material products they provide, they have been highly valued as pets and protectors. The latter is especially true of white-fronted geese. See Vonga Bobby's account of how greater white-fronted geese saved a women and children from a brown bear (page 39). Geese have also been trained to sight game for hunters.

A traditional story explains the reason for the snow goose's black wing tips: The raven and the snow goose wanted to get married. When migration time came, the snow goose held on to the raven to assist him on his unfamiliar southern journey. The raven, unable to accomplish it, left black marks on the snow goose's wing tips with which she held him.

WATERFOWL

Ducks: *Jija*

At least twenty species of duck migrate to the Lime Village area, the first arriving by mid-April after the swans and geese when open water first appears along lake shores and at creek mouths. Normally the majority of ducks have returned by late April, scoters being the last arrivals. Some ducks breed in the area, mostly in the lake country, while others merely rest before migrating elsewhere. By mid-October most summer resident and migrant ducks have returned south. Lime Villagers report that the number of ducks in the area has decreased in recent time.

Lime Villagers have usually harvested more ducks during a year than other waterbirds. They are snared and shot with bows and arrows and guns.

Like other waterfowl, ducks have provided essential food as well as a welcome change from a winter diet. Preferred ducks are those with the highest fat content, although all ducks are said to be relatively fat in the spring and fall. During the nesting season, waterfowl lose fat although scoters, bufflehead, and goldeneyes remain relatively fat all the time. Given an equal chance among ducks, hunters have tended to shoot the largest, high fat content ducks. Often they have not had this option and harvest, often due to necessity, the most available, edible duck. If a person is very hungry, he may eat the raw outer portion of a duck's gizzard.

Lime Villagers have employed duck skin and feathers for clothing and bedding, hunting equipment, and decorative items as they have many other birds. See northern shoveler for additional information regarding duck-bill spoons.

Dabbling Ducks

Dabbling ducks are those ducks that remain on the surface and either feed there or dip up and feed on vegetation under water.

WOOD DUCK (*AIX SPONSA*)

generally recognized but not recorded

Description: Males are unmistakable with a crest of glossy, green feathers, black and white face, red base of bill, reddish breast, beige sides, and glossy bluish back. Females have a partial crest and white patch around the eye. The wing speculum is blue.

People report that wood ducks have bred in the area. Lime Villagers were certain this bird was seen there. Wood ducks are casual spring and summer visitants to southeastern Alaska from Ketchikan to Juneau, and have also been reported on Kodiak Island in 1994 (West 2002). It is conceivable that it might also be found on the Stony River.

GADWALL (*ANAS STREPERA*)

generally recognized but not recorded

Description: The male is a gray-brown duck with a dusky breast, pale brown head, and black at the base of the tail. The belly is white. There is a chestnut patch on the forewing and a black and white speculum. The female is brown with a lighter head, white belly, and the same wing pattern as the male.

The gadwall is a very rare visitor to Interior Alaska. Many hunters misidentify this duck, especially females, because they lack easily recognized characteristics. Gadwalls have been used as described for other ducks.

AMERICAN WIGEON (ANAS AMERICANA), "WHISTLER"

sheshinya (name refers to its vocal sound)

Description: Males have a white cap and forehead, gray head with an often inconspicuous green stripe behind the eye, a pinkish breast and sides, and brown back. Females have an all gray head, but retain the pinkish breast and sides of the male. Both sexes have a white forewing patch and a green wing speculum. Probably the Dena'ina name refers to the male's call, a three-noted whistle with a higher pitched second note.

Lime Villagers observe that the wigeon that breeds in the area often flies in large flocks. Although it prefers lakes, it sometimes occurs in small streams. Besides hunting the American wigeon for food, Lime Villagers have used it for clothing and bedding.

MALLARD (ANAS PLATYRHYNCHOS)

chadatl'ech'i "the one with a blue head"

Description: Males have a glossy iridescent green head, yellow bill, brown breast, gray back, orange feet, and small upper tail coverts that curl upwards. The wing speculum is bright blue bordered by white. Females are mottled brown all over, but have orange feet and the blue speculum like males. The female gives the characteristic "quack."

Mallards nest in marshes in the area and nests have been sighted on dry tundra near trees. Mallards are said to have especially good  eyes making them difficult to surprise when hunting them. They also fly straight up when taking off.

The curled tail feathers are an important cultural feature of the mallard because a traditional story describes their earlier use.

Although most mallards in the Lime Village area are migrants, a few birds winter in small streams and lake outlets that do not freeze. They and other species that missed the fall migration are often able to survive in these locations until summer.

Mallards and other waterfowl have provided food not only in the spring and fall, but the birds that stay throughout the winter have supplied emergency food. Some people say that they are fat and taste good, while others dislike their "fishy" taste due to their eating fish eggs. Clothing and bedding have been made from mallards as from other waterfowl. A Lime Village elder remembers a beautiful feathered neck skin mallard jacket. Lime Villagers have crafted decorative baskets from the dried, green-feathered skin portion of the mallard's head.

WATERFOWL

NORTHERN PINTAIL (ANAS ACUTA)

tsendghinlggesh "it walks down toward the water"

Description: Males are characterized by a brown head, a long slender white neck, white breast, a thin white stripe extending up the side of the head, a gray body, and a long pointed tail. Females are mottled brown but have the same long-neck proportions as the male.

The northern pintail breeds in area lakes. Besides being hunted for food, the northern pintail's skin and feathers have provided clothing and bedding. The upper Stony River people observe that pintails are fast flyers and are "smart" about staying out of shooting range.

NORTHERN SHOVELER (ANAS CLYPEATA), "SPOONBILL"

veduzhizha dghiłtali "the one whose bill is wide"
vedushqula "its bill (is a) spoon"

Description: Males have a glossy green head, white breast, reddish brown sides, and a large spatulate bill. Females are mottled brown but are identified by the large spoon shaped bill. There is a pale blue forewing patch present in both sexes.

People have found shoveler nests along the sides of small streams that run into lakes. Besides using the shoveler for food, clothing, and bedding, Lime Villagers have prized shoveler bills for spoons more highly than other bird bills. The especially wide bill is boiled until it is soft and its skin can be removed. The bill may then be tied on to a wooden handle.

GREEN-WINGED TEAL *(ANAS CRECCA)*, "POCKET DUCK"

qulchixa "the one that bounces up"

Description: Males have a brown head with a bright green stripe back of the eye, a gray body, and a green wing speculum. Females are mottled brown but also have the green speculum. The green-winged teal, as its local name "pocket duck" suggests, is one of the smallest ducks in Alaska. The term *qulchixa* is not fully analyzable but in other dialects its Dena'ina name means "the one that flies up," referring to the teal's ability to fly straight up from the water.

The green-winged teal occurs in the area's lakes and small streams where it nests in vegetation immediately above the water's edge. They are often seen in pairs and commonly pop up in front of boats. People also observe that they walk easily on the ground and like to perch on logs and willow branches. Lime Villagers have hunted green-winged teal for food, clothing, and bedding materials. Because of its small size, it has not been as intensively sought after as many of the larger ducks.

WATERFOWL

Diving Ducks

Diving ducks are those that dive under water to pursue fish, aquatic insects, mollusks, or crustaceans.

CANVASBACK (*AYTHYA VALISINERIA*)

veq'es dasdeli "the one whose neck is red"

Description: Males have a long black bill, dark reddish head and neck, black breast and tail, and white back and belly. Females have a light brown head and neck and gray body, but the sloping forehead to the long dark bill is distinctive.

The canvasback breeds in lakes in the Lime Village area where it is quite abundant. Although it has been hunted for food, elders say that the duck is a fairly recent arrival in the area, and they do not remember feathered-skin clothing being made from it.

REDHEAD (*AYTHYA AMERICANA*)

generally recognized but not recorded

Description: Male has a dark red head, black breast and tail, and gray back and sides. The bill is bluish gray with a dark tip. Differs from similar canvasback by its rounded head and grayer back. Female is dark brown with no conspicuous marks. There is no wing patch evident in flight in either sex. Redheads are rare in the Lime Village area. The Dena'ina may group this duck with the canvasback as they are very similar in appearance.

RING-NECKED DUCK (*AYTHYA COLLARIS*)

generally recognized but not recorded

Description: Similar to the scaup, the male ring-necked duck has a darker back with a vertical white streak between the black breast and dark shoulder. The gray bill has a white ring around it. The neck ring is brown and difficult to see. Females are all brown with a light eye ring, light line back of the eye, and lighter face. The bill also has a white ring. Ring-necked ducks are rare visitors in the Lime Village area. The Dena'ina may consider this to be a type of scaup.

GREATER SCAUP (*AYTHYA MARILA*), "BLUEBILL"

vech'enlna q'enk'elggeyi "the one with white on its wings"

Description: Male has black (usually iridescent green) head, black breast and tail, and white body. The back appears gray. The bill is blue (scaup are often called "bluebills"). Female is all brown with a white crescent at the base of the bill. The white wing stripe in both sexes extends from the base of the wing to the long primary flight feathers.

GREATER SCAUP

LESSER SCAUP

LESSER SCAUP *(AYTHYA AFFINIS)*, "BLUEBILL"

vech'enlna q'enk'elggeyi "the one with white on its wings"

Description: Male has black (usually iridescent purple) head, black breast and tail, and white body. The back appears gray. The bill is blue. Female is all brown with a white crescent at the base of the bill. The white wing stripe in both sexes extends from the base of the wing to the bend in the wing and does not include any of the long primary flight feathers.

The scaups have been used as described for other ducks.

KING EIDER *(SOMATERIA SPECTABILIS)*

generally recognized but not recorded

Description: Adult males have black back, sides, and belly; the breast is white. The head has a bluish crest or helmet. The bill is pink and there is a yellow shield over the base of the bill. Immature males have dusky brown heads, brown back and belly, and white breast. The bill is pale. Females are all brown, sometimes reddish-brown mottled with dark brown. The head is paler and the elongated bill is horn color.

A king eider was reported below a canyon area above Lime Village on the Stony River. Presence of a king eider away from the coast is accidental. A Dena'ina term for the common eider, *qaniłghach (SOMATERIA MOLLISSIMA),* has been recorded in Outer Inlet and in Iliamna

78

HARLEQUIN DUCK *(HISTRIONICUS HISTRIONICUS)*, "STONE DUCK"

tus qet'ay "resident of the passes"

Description: Males are brightly-colored dark ducks with white spots and patches and rusty sides. Females have some of the same white spots but are generally dusky gray in color. They inhabit fast flowing rocky streams of the upper Stony River area. Reasons for its local English name are the fact that it lives in a rocky environment and its habit of standing on stones in streams. Nests were found in river banks.

People have harvested harlequin ducks for food and for their colorful feathers prized for traditional dance headwear. Their feathered skin and down feathers have provided clothing and bedding.

An elder comments that harlequin ducks are difficult to harvest because they spread out in the creeks and do not congregate in large numbers like many other ducks. A younger person observes that, once found, they are easy to shoot because they are not quick to fly. Because of their high fat content, they have been especially enjoyed for food.

WATERFOWL

Scoters, "Black Ducks": *Jijalt'esha "black duck"*

Surf Scoter *(Melanitta perspicillata)*

venchix va'ilch'eli "the one with a light-color on its nose"

Description: Male is all black with a large white patch on the nape and a white forehead. The bill is red orange at the tip and white at the base (nose color referred to in Dena'ina name). The eye is white. Female is all dark except for two white spots on the face.

White-winged Scoter *(Melanitta fusca)*

venaq'a qa'ilch'eli "the one with light-colored eyes"

Description: Male is all black but has a white patch behind the eye and a white eye as the Dena'ina recognize, a reddish or orange bill, and a large white wing patch. Female is all dark except for light spots on the head and the large white wing patch.

Black Scoter *(Melanitta nigra)*

venchix va'ldetsiggi "the one with yellow-orange on its nose"

Description: Male is all black with an orange or yellow bulb at the base of the bill as the Dena'ina name indicates. Female is all dark except for a light tan face.

BLACK

WHITE-WINGED

SURF

MALE SCOTERS

All three species of scoter breed in the upper Stony River area. Most Lime Villagers prefer scoters for food over other ducks because they consistently have a thicker layer of fat than other ducks in the area except perhaps the bufflehead. However, because they are larger and more abundant than the bufflehead, they have been harvested more frequently. Wearing a scoter hat is said to prevent early gray hair.

Scoters are especially slow at taking off. Like other diving ducks, they take off into the wind. Because they are black and frequent deep water lakes, scoters are easily camouflaged especially by waves on a windy day. However, they often go to protected lake and stream bays when it is very windy. Because of these characteristics, hunters have made greater effort than with other ducks to have the sun at their back and to be upwind from the birds especially when the birds take off. Ideally, they prefer to hunt them on calm days. If a scoter's partner is killed, it will often circle around the area looking for it and can be lured towards a hunter who splashes the water with a pole.

WATERFOWL

LONG-TAILED DUCK *(CLANGULA HYEMALIS)*
(formerly named Oldsquaw)

ahhanya name refers to its vocal sound

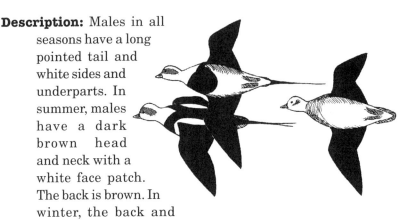

Description: Males in all seasons have a long pointed tail and white sides and underparts. In summer, males have a dark brown head and neck with a white face patch. The back is brown. In winter, the back and head molt to white except for a brown patch at the "ear." Females are mottled brown on the back, white on the sides and below. The head is browner in summer than in winter, but there are usually one or more brown patches on the light head.

The long-tailed duck is an uncommon migrant in the Lime Village area.

BUFFLEHEAD *(BUCEPHALA ALBEOLA)*, "BUTTERBALL"

sukna tsighał "wool fabric hairnet"

Description: Male's head is usually puffed up and has a large white patch, a possible reason for the meaning of its Dena'ina name. The back is dark, the breast and belly are white. There is a large white wing patch. Female is generally brown but has a white patch on the side of the head and the white wing patch. The bufflehead, the smallest North American duck, summers in the upper Stony River area and is found near banks of small streams.

Lime Villagers prefer the bufflehead for food over many other ducks because of its high fat content. The bird has also been hunted for clothing and bedding. People observe that buffleheads are fast moving birds and that a hunter needs to aim quickly especially when birds are flying towards him.

Common Goldeneye *(Bucephala clangula)*, "Whistle Wings"

tsiq'unya "ridged head"

Description: Male has a round white patch below the bright yellow eye, and a distinctively shaped, greenish-black head. The unusual shape of the duck's head is suggested in the meaning of its Dena'ina name. The dark back and tail contrasts with the white wing feathers and white belly. Female has a dark brown head, gray body, white wing patch, and a dark bill with a yellow tip. Both make a loud whistling with the wings in flight.

The common goldeneye breeds in the area and nests in hollow trees near water. It and Barrow's goldeneye have been favored for food because of their relatively thick fat layer compared to most other ducks. Hunters observe that due to their speed, it is hard to kill a goldeneye and that they don't die easily unless hit in the head.

BARROW'S GOLDENEYE (*BUCEPHALA ISLANDICA*), "WHISTLE WINGS"

tsiq'unya "ridged head"

Description: Male similar to common goldeneye, except the white patch at the base of the bill is crescent shaped, and the head has a purple shine instead of green. Females are very similar to the common goldeneye, but the bill is almost all yellow.

Barrow's goldeneye is less common in the area than the common goldeneye. Uses for this species are the same as described above.

Mergansers: "Fish Ducks"

Mergansers are ducks that have a relatively long and narrow bill that is serrated or toothed along its margin. They swim underwater and catch fish grasping them with the toothed bill.

COMMON MERGANSER (*MERGUS MERGANSER*), "FISH DUCK"

cheghesh √

Description: Male has green head, red bill, and white body. Female has reddish brown head, white throat, reddish bill, and gray body. Both sexes have a white wing patch visible in flight.

RED-BREASTED MERGANSER *(MERGUS SERRATOR)*, "FISH DUCK"

yucheghesh "sky merganser"

Description: Male has crested green head, red bill, pale rusty breast, black and white patterned body, and gray sides. Female has reddish brown head, light throat, gray body, very similar to common merganser. Both sexes have a white wing patch visible in flight.

Mergansers summer in the streams of the upper Stony River area. The red-breasted merganser leaves in September with the majority of other ducks, while the common merganser remains until October.

Although the upper Stony River people have consumed mergansers, they do not prefer them for food because of their fishy taste. The eggs have not been harvested except in emergencies.

Due to the common merganser's swiftness, Raven placed a carved image of the duck on his boat in the traditional story "Raven Rescues His Wife" told by Alexie Evan (Tenenbaum and McGary 1984, 109).

Osprey, Eagles, Hawks, Hawk-Like Birds, & Owls

The Dena'ina classify the osprey and eagles separately from the other hawks and falcons. Identification of the hawks of the area has been more difficult than for most other groups. Thus this category may contain misidentified species due to lack of information about specific birds.

Lime Villagers have especially harvested the larger hawks and falcons for their skin and feathers to make clothing and bedding and used their feather shafts in constructing snares. While hawks and falcons appear to have only been used as emergency food, some of the larger species have been tamed as pets and hunting assistants.

Most hawk-like birds are migrants, normally leaving by December when the weather becomes extremely cold and returning in the spring. The Lime Village name for April translates as "soaring hawk month" (J. Kari 1977, 144).

Osprey

OSPREY (*PANDION HALIAETUS*), "FISH HAWK"

tahch'ek'an "the one that watches the water"

Description: Sexes similar. A large, long-winged hawk, brown above and white below. The head is white with a short shaggy crest and a dark stripe back of the eye. Ospreys fly with their wings bent back.

The osprey feeds almost exclusively on live fish and often perches above the water or hovers over the water searching for fish near the surface. The meaning of its Dena'ina name "one that watches the water" suggests this habit. The osprey occurs in the Lime Village area throughout the open water season, leaving when it is no longer able to fish.

The osprey has provided feather-skin clothing and other material items (see Bird Clothing).

Lime Villagers have tamed young ospreys both for pets and hunting assistants. An osprey can be trained to bring fish and rabbits to its tamer and to help find game.

A Lime Village resident remembers taming an osprey for a pet that she kept at a fish camp. It was taken from its nest in June and flew away in September when the geese left. The bird was fed mostly fish but also gray jays, red squirrels, mice, grouse, hares, and meat scraps. When the bird was old enough, it sometimes flew all day alone and hunted for itself but would return at night. The osprey, which had a pleasant disposition, was greatly enjoyed by the camp residents and was given a personal name.

A traditional upper Stony River Dena'ina belief is that if a wild osprey flies by a person and drops something, it foretells misfortune. Apparently no wild bird that drops something near a person signals good luck.

Eagles

Bald Eagle (*Haliaeetus leucocephalus*)

ndalika'a "big fliers"

Description: Adults are large long winged hawks with dark brown body and wings and white head and tail. The bill and feet are yellow. Eagles take five years to achieve full adult plumage. Immatures have blotchy white markings on the belly, under the wings and at the base of the tail. The bill is gray and lightens with age. Bald eagles often soar with the wings held flat rather than in a *V*.

GOLDEN EAGLE *(AQUILA CHRYSAETOS)*, "MOUNTAIN EAGLE"

yudi √

Description: Adults are large all brown long-winged hawks with a faintly banded tail. The head is usually paler brown than the body. The bill is horn color and the feet are yellow. Immature golden eagles show distinct, not blotchy, white patches at the base of the primaries and base of the tail. Golden eagles usually soar with the wings held in a shallow *V.*

The bald eagle and the golden eagle breed in the upper Stony River area. They normally arrive in March and migrate south in December, but if the weather does not become extremely cold, they may remain throughout the winter. In January 1992, Lime Villagers saw a bald eagle in the area. The bald eagle usually resides in forests near bodies of water while the golden eagle is largely a mountain inhabitant. Three Dena'ina place names in the upper Stony River area around Kristin Creek refer to the golden eagle, e.g. Yududuhtnu "golden eagle stays-creek" (P. R. Kari 1983, Map 3).

People emphasize the strength of the golden eagle. An elder remembers when a golden eagle took a three-foot brown bear cub from its mother, then smashed it by flying high and dropping it. When the elder's father told him to take the cub from the eagle, the eagle did not attempt to stop him. They observed that the eagle's claws went through the cub's heart. People have protected young children from becoming potential golden eagle victims.

OSPREY, EAGLES, HAWKS, HAWK-LIKE BIRDS, & OWLS

Eagles have been hunted with bows and arrows, snares, and with guns. Snares made from feather shafts, braided spruce roots, and braided caribou or moose sinew were placed where eagles tended to land and were baited with fish or meat. An elder explains that they were harvested only when they were really needed. Younger hunters explain that they do not kill eagles when the birds attempt to take their freshly harvested ducks because their elders told them not to kill eagles.

In times of food shortage, Lime Villagers have harvested eagles, but using them for food has not been a common practice. One fall when food was scarce, an elder remembers eating a bald eagle that was so fat it couldn't fly. The person singed the eagle over a fire like waterfowl and said that although it tasted like fish, it was good and that someone made bread with the grease.

People have valued the feathered skin of both eagles for making strong, warm clothing and bedding (cf. P. R. Kari 1983, 6; see also Bird Clothing). Eagle feather shafts, said to be the strongest of all feather shafts, have been used in crafting various technological items (see Feather Technology). Eagle feathers have been the most highly esteemed feathers used in headbands. An elder explains that feather headdresses were worn by the singers at potlatches and that a chief did not wear them unless he sang at the gathering. Shamans also wore headdresses (J. Kari 1977, 239). The band itself was made of tanned bear skin to which upright feathers were sewn. Bald eagle and golden eagle feathers could be mixed or worn separately on a head band. They were valued equally although golden eagle feathers are harder to obtain than bald eagle feathers because golden eagles, largely mountain dwellers, generally stay further from human habitation than bald eagles.

Eagles have apparently been trained to assist hunters. Hunters have learned the location of waterfowl in wetlands by watching where wild golden eagles land.

Traditionally certain children have worn a part of a golden eagle on their clothing as an amulet to help them become good hunters. They were told never to kill a golden eagle.

Hawks & Falcons

NORTHERN HARRIER (*CIRCUS CYANEUS*)
(formerly named Marsh Hawk)

> *k'kakenk'delkidza* "the one with a pattern on the base of its tail"

Description: Both sexes may be recognized by the white rump patch, the long tail, thin wings, and their habit of soaring low over open country hunting for small mammals and birds. Male is gray above  and light below; the wing tips are black. Female is brown above and often streaked with orange-brown below. Juveniles resemble the female.

The northern harrier is most commonly seen in the open country near waterfowl harvesting areas. See Tenenbaum and McGary 1984, 50–57 for the traditional story "K'kaken'delkidza Marsh Hawk" as told by Alexie Evan.

Accipiters

Accipiters or woodland hawks have short rounded wings and a long tail. They fly through the woods hunting for small birds that they can pursue through branches to catch them.

SHARP-SHINNED HAWK *(ACCIPITER STRIATUS)*, "CHICKEN HAWK"

generally recognized but not recorded

Description: Identified by small size, rounded wings, and long tail. Adult is gray above and light below, often with fine orange barring. Immatures and birds in their first year are brown above and have varying amounts of brown streaking on the breast and belly. The sharp-shinned hawk, a permanent forest resident of the upper Stony River area, hunts red squirrels, other rodents, small birds, and possibly grouse. It has been known to overwinter in the area.

NORTHERN GOSHAWK *(ACCIPITER VELOX)*, "CHICKEN HAWK"

gizha kegh "large gray jay"

Description: Large hawk with rounded wings and long tail, often hunting in deep woods for grouse, hares, small birds, and rodents. Adult is blue gray above and lightly barred with gray below. There is a distinct dark line through the eye and a light line over the eye. Immatures and first year birds are brown above and streaked below. The generally gray plumage is referred to in the Dena'ina name. This bird is linked by name with two other birds: *gizha* gray jay or camp robber (though this term is not used as such in most of Dena'ina) and *gizha vay* northern shrike.

94

The goshawk is apparently the only regular permanent resident hawk. It receives its local English name from its habit of eating grouse.

A goshawk claw worn by a child is said to help him become a good hunter (J. Kari 1977, 240).

Soaring Hawks *(Buteo* sp.)*: Q'uluq'eya*
"the one that soars around"

Buteos are large hawks with long wings rounded at the tips and a broad tail. As the Dena'ina name indicates, they soar in large circles high over open areas searching for small mammals below. In Dena'ina the group of hawks called *q'uluq'eya* has three members, one of which is the peregrine falcon, which is not a Buteo. One place name on the lower Stony River is for the soaring hawks: Q'uluq'eya Tuyana, "soaring hawk straight-stretch."

RED-TAILED HAWK *(BUTEO JAMAICENSIS)* AND/OR HARLAN'S HAWK *(BUTEO JAMAICENSIS HARLANI)*

ch'vala q'uluq'eya "spruce soarer"

Description: Harlan's hawk is a subspecies of red-tailed hawk. Both red-tailed and Harlan's hawks have light and dark phases called morphs. The dark morph Harlan's hawk is the commonest variation in the upper Stony River area spruce forests. The body plumage is all dark, almost black in some individuals, the tail is lighter sometimes with a tinge of reddish-brown towards the tip. They soar over woods and open country on broad rounded wings. Red-tailed hawks have a distinct solid red-brown tail. Both dark and light phases of Harlan's hawk occur in the Stony River area.

According to Lime Villagers' observations, the dark Harlan's hawk is more prevalent in the area than the red-tailed hawk. However, a red-tailed hawk nest was identified on a cliff near a spruce forest.

ROUGH-LEGGED HAWK (*BUTEO LAGOPUS*)

dghiliq' q'uluq'eya "mountain soarer"

Description: Identified by the long white tail with dark band at the tip. Some individuals are very light with white wing linings and brown back and brown streaks below. Other birds are almost black above and below, but have the light tail. Rough-legs have relatively long wings and often hover high over open country where they search for small mammals.

The rough-legged hawk nests in Lime Village area cliffs located near or in the mountains. The location of the nest and its habit of flying high in mountainous areas are reasons for its Dena'ina name. The bird has apparently also been seen in other habitats such as along lowland streams.

Falcons

Falcons are identified by their long pointed wings and long tail. They are swift fliers and usually dive at their prey of small birds or mammals often from great height. They appear to have small heads and large eyes.

AMERICAN KESTREL *(FALCO SPARVERIUS)*, "SPARROW HAWK"

generally recognized but not recorded

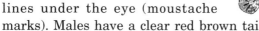

Description: Kestrels are small falcons that often hover while searching for their prey of small birds or insects. They are birds of open fields or forest edge. Both sexes are generally reddish brown above and have a reddish tail. They have dark vertical lines under the eye (moustache marks). Males have a clear red brown tail and bluish gray wings that the female lacks. The kestrel is probably very rare in the Lime Village area.

MERLIN *(FALCO COLUMBARIUS)*

k'enchix t'it'a "the one with a twisted nose"

Description: Merlins are small falcons that usually hunt in woods or mixed open woodlands. Both sexes are streaked below and dark above; the male is bluish-gray above, the female brown. The tail is heavily banded.

Merlin nests have apparently been found on cliff ledges in the area. This small falcon is known for preying on birds.

OSPREY, EAGLES, HAWKS, HAWK-LIKE BIRDS, & OWLS

Gyrfalcon *(Falco rusticolus)*, "Bullet Hawk"

qenay, qennay "?"

Description: The gyrfalcon has broader wings and is larger and heavier appearing than the peregrine falcon. Color variations occur from birds that are almost white with dark barring to almost black. Most individuals are intermediate with blue-gray back and dark streaks on light undersides. The wings and tail are barred.

The gyrfalcon nests on steep mountain sides. Reportedly it can kill a person close to its nest by forcefully hitting him. A Lime Village resident observes that the gyrfalcon cracks bones to obtain their marrow by dropping them from mountain cliffs. This falcon is likely a permanent resident of the area.

Peregrine Falcon *(Falco peregrinus)*, "Duck Hawk"

tsanenh q'uluq'eya "cliff soarer"

Description: Peregrines are large powerful falcons that are very fast fliers and kill ducks and other birds in the air by a blow with the feet. Adults are dark blue-black above and barred below; immature and first year birds are dark brown above and streaked below. The tail is barred. The wings are held bent at the wrist and the head appears tucked back into the body giving the appearance of no neck. They often nest on cliff faces.

People observe that the peregrine falcon flies along streams and lakes hunting ducks.

Hawk-Like Birds

NORTHERN SHRIKE (*LANIUS EXCUBITOR*)

gizha vay "gray gray jay"

Description: A large, pale-gray bird with white underparts, dark wings and tail, and a black mask. The bill is large and hooked. There are conspicuous white wing patches and white on the outer tail feathers. The shrike is a permanent resident of the upper Stony River area. Being a hunter without the talons of hawks and owls, it kills or stuns mice and flying birds with its strong beak, then hangs them from a tree crotch or pierces them with a twig. Some Dena'ina relate the shrike to the hawks because of its habits and hooked bill, as well as by the term *gizha* that appears in the term for goshawk *gizha kegh*.

Lime Villagers have made blankets from the feathered-skin of northern shrikes. The birds are not used for food except in an emergency.

A traditional belief is that if a shrike goes near people, they will have good luck. Because the shrike is said to dry and store its food like people, the luck may refer to good harvesting success.

Owls: *K'ijeghi*

While most owls are permanent residents, the short-eared owl is a migrant and snowy owls often come south in winters when their prey is scarce in the north. Most owls hunt at dusk or dawn, but the snowy, short-eared, and northern hawk owls are day hunters.

OSPREY, EAGLES, HAWKS, HAWK-LIKE BIRDS, & OWLS

Owls are especially accredited with predicting upcoming major events, particularly misfortunes. It is important to remember that they do not cause misfortune but merely foretell its occurrence. Although certain other birds also have this ability, it seems to be most highly developed in the owls. The number of accounts of owls warning humans of misfortune, especially of death, is significantly greater than for other birds. If an owl, especially a great horned owl calls a person's name, it means that person may die. An owl should never be interrupted when speaking.

Owls making unusual sounds or exhibiting unusual behavior can also be omens of misfortune. Unusual behavior includes flying close to a person's house, staying near the house, and touching a person or his clothing. People inform one another of significant owl behavior.

For a change of diet, the large owls, said to taste something like turkey, have been hunted for food, but owl eggs have only been harvested if other food was scarce.

The skin and feathers of the larger owls have been made into clothing including coats, hats, pants, and mittens. Owl claws also served as amulets.

Although occasionally tamed, owls have not been favored as pets because of their mean disposition. An owl kept at a fish camp was built a "house" from which he occasionally flew to catch mice to supplement the fish it was fed. The owl apparently did not hunt as much by itself as hawks kept at the camp in other years. Apparently owls have not actively assisted people in hunting game as ravens and hawks, although one species is known to nod toward a bear den.

GREAT HORNED OWL (*BUBO VIRGINIANUS*)

k'ijeghi "the eared one"

Description: A large, heavily barred and streaked brown owl with conspicuous ear tufts as noted by the Dena'ina name. The chin is white. Sexes are alike. Great horned owls are common in woods throughout Alaska.

Both cultural and linguistic evidence support the conclusion

that the great horned owl is the most important owl species. In the Inland Dena'ina language, the general name for owl and the great horned owl's name are the same, an indication of the significance of the great horned owl. For example, its communication abilities with people, a specialized trait of owls, is said to be more developed than that of other owls. All owls are said to be able to talk to people in whatever language necessary and to warn them of negative events, but the great horned owl is especially adept in these skills. For example, the great horned owl is known for predicting bad weather such as snow in winter or rain in summer. See below for other examples.

People have taken young great horned owls from nests and trained them as pets.

SNOWY OWL (*NYCTEA SCANDIACA*)

yesvu "snow white"

Description: Adult snowy owls are white (as in the Dena'ina name) with some dark barred markings on the body, wings, and tail. Juveniles are more heavily barred with brown. They have no ear tufts. Sexes are alike. Snowy owls are arctic tundra birds and come south when their diet of lemmings and voles becomes scarce in arctic Alaska. They prefer open areas as along the coast and in alpine tundra and occur rarely in the upper Stony River area.

The snowy owl is apparently a good luck sign (J. Kari 1977, 238).

NORTHERN HAWK OWL *(SURNIA ULULA)*

delukdiday "one that sits on branches"

Description: A medium-sized owl with a long tail that often perches on a high tree branch as its Dena'ina name suggests. The undersides are heavily barred with brown and the back is speckled with white. There are no ear tufts. The sexes are alike. Hawk owls, permanent residents of the Lime Village area, hunt during daylight hours for small mammals and birds.

Young hawk owls have been taken from nests and raised as pets. If someone lies, the hawk owl is said to sing a special song that other types of birds listen to and can understand. The owl also dances merely for "fun" and likely for other reasons.

An elder relates that one year there was a lot of rain in the winter that occurred soon after the hawk owl "hollered" in a certain way. The call is *dugu dugu*.

GREAT GRAY OWL *(STRIX NEBULOSA)*

nułdes "?"

Description: The largest owl in the Lime Village area, the great gray is speckled and barred gray, brown, and black. It has no ear tufts, but a very large round head with concentric rings of barred feathers around the eyes. Sexes are alike. It is well camouflaged when it sits motionless in deep woods by a spruce trunk. It preys on small mammals throughout the year in the woods around Lime Village.

SHORT-EARED OWL *(ASIO FLAMMEUS)*

naghuk'elts'eha
"the one licking
something for us"

Description: A buffy, brown owl that flies over open country during the day in search of small rodents. Sexes are alike. Its moth-like flight where it wavers and dips and dives as it hunts is characteristic. Most short-eared owls are migratory although occasionally one will winter in the area.

BOREAL OWL *(AEGOLIUS FUNEREUS)*

skitnaz'una "the one that stays under trees"

Description: The boreal owl is a very small owl, has a white facial disk bordered by black, and a pale bill. The forehead is speckled with white. Sexes are alike. It hunts for small mammals and birds at night and nests in a hollow tree. It is a permanent resident of the Lime Village area. The northern saw-whet owl (*Aegolius acadicus*) would be an accidental visitor to the Lime Village area and, if so, may be identified by the same name.

Although not as significant as the great horned owl, the boreal owl appears to be an especially important owl. According to a traditional Dena'ina belief, if a boreal owl makes a noise near a house, a household member will catch an animal the following day. However, if no animal is caught, bad luck will occur. The owl doesn't cause misfortune, but due to its actions merely gives warning.

A person remembers that his father warned him not to climb a tree where a boreal owl nested because the owl is strong enough to knock a person from the tree.

People observe that they can touch the resting bird during the day but that it is "wild" at night. While it is not unlucky for a person to touch the owl, misfortune may occur if the bird hits a window.

OSPREY, EAGLES, HAWKS, HAWK-LIKE BIRDS, & OWLS

Grouse & Ptarmigan

Grouse and ptarmigan are chicken-sized birds that spend most of their time on the ground. Their flight is rapid and direct; they usually jump from hiding and fly swiftly in a whir of wings. The fan-shaped tail, which is often used in courtship display, is an important identifying feature.

Ruffed Grouse (*Bonasa umbellus*), "Willow Chicken"

k'delneni "the one that pounds"

Description: Males are brown or gray (two color phases), have a conspicuous dark ruff at the neck that is puffed out during courtship, a small crest, and a finely barred tail with a heavy dark band and thin white terminal band. Females are similar but lack the ruff and crest. Male ruffed grouse display by rapidly beating their wings while sitting on a favorite log or other perch. The air trapped under the beating wings creates a series of accelerating thump sounds called drumming. As the Dena'ina name reflects, Lime Villagers have observed ruffed grouse drumming on logs and on gravel. Ruffed grouse live in deciduous and mixed woods and prefer berries, insects, and seeds.

Spruce Grouse (*Dendragapus canadensis*), "Spruce Chicken"

eldyin "the one that eats spruce boughs"

Description: The male is mostly gray, black, and white with a red comb over the eye, that is most visible during spring. The tail is dark with a brown tip; the undertail coverts are spotted with white. Females are browner than males and the tail is barred, but the tip remains brown. Spruce grouse are often easily approached and females on the nest can sometimes be caught by hand. They often sit in spruce trees where they feed on needles in winter, thus the Dena'ina name.

106

WILLOW PTARMIGAN *(LAGOPUS LAGOPUS)*

q'ach'ema "?"

Description: In winter, birds are all white with a black bill, eye, and tail terminating in a narrow white band. The feet are heavily feathered. In spring, males develop a red comb over each eye and the head and neck molt to a chestnut brown. In summer, both males and females are mottled brown, the female duller than the redder brown of the male. The wing feathers are white all year. The principal diet of the willow ptarmigan consists of buds and twigs of willow shrubs, occasionally dwarf birch, and other small shrubs. In summer and fall, they also eat berries and insects.

ROCK PTARMIGAN *(LAGOPUS MUTUS)*

jeł q'ach'ema "mountain ?"

Description: In winter, birds are all white with black bill, eye, and tail. The male has a black line through the eye. Females are distinguished from similar willow ptarmigan by their smaller size, especially the bill. In spring, males develop bright red combs over each eye. Summer birds are mottled brown with white wings. Rock ptarmigan eat buds, twigs, leaves, and berries of smaller shrubs, including alder.

WHITE-TAILED PTARMIGAN (*LAGOPUS LEUCURUS*)

qatsinɫggat "you are dreaming," also called *ch'etl'
q'ach'ema* "willow ?"

Description: In winter, birds are all white showing only a black
eye and bill; the tail is white. Summer birds are mottled brown
with white wings and tail.

All three species of ptarmigan live in the upper Stony
River area. White-tailed
ptarmigan are exclusively
upland birds while the willow ptarmigan, and to a
lesser extent the rock ptarmigan, move to the lowlands during high
winds and other extreme winter conditions in the mountains. They
find protection and food in the stream valleys. They return to the
mountains to mate in late March or April.

Because the willow ptarmigan is the most easily accessible of the
three species, Lime Village people have hunted willow ptarmigan
more frequently than the other ptarmigan. Another indication of
the special importance of the willow ptarmigan is that its Dena'ina
name *q'ach'ema* is also the general name for ptarmigan. The primary hunting season for ptarmigan is approximately November
through March.

Traditionally ptarmigan have been taken with bow and arrows and
with snares, often in fence-like piles of brush, called *heɫ* (see Blinds
for a description of brush fences.) People have set trout nets in
willow groves and chased ptarmigan toward the net where the birds
become entangled and are easily killed. Hunters have snuck up
behind ptarmigan on the ground and simply grabbed them with
their bare hands.

While the most common method of preparing ptarmigan for food is
boiling, they are also roasted on a stick over a campfire. The entire
bird is eaten except for the feathers, bones, crop, feet, and intestines, the latter being too small and thus too time-consuming to
clean. Like spruce grouse, ptarmigan can be consumed raw or partially raw.

Although ptarmigan feathers have been used as bedding and clothing filling, their skin is too thin for making feathered-skin garments. Children have worn dried ptarmigan feet on their shoes to make them fast runners.

In the traditional story *Raven Rescues His Wife* told by Alexie Evan (Tenenbaum and McGary 1984, 109), the raven tells the ptarmigan and spruce hen people to sew a skin boat that he will use to rescue his wife. Young girls have worn a spruce hen foot as an amulet to help them become good seamstresses (J. Kari 1977, 240).

SHARP-TAILED GROUSE *(TYMPANUCHUS PHASIANELLUS)*

k'elteli "the one that makes a thumping noise"

Description: Sharp-tailed grouse are birds of open country but in the Lime Village area they are found in both closed and open deciduous and spruce forests, and on tundra and other open country margins. They are speckled brown above and below and blend in with dry grass. The central tail feathers are elongated and the margins of the tail, visible in flight, are white.

The spruce grouse and ruffed grouse are the most common grouse of the upper Stony River area. The sharp-tailed grouse appears to be rare. In the 1970s an elder observed that flocks would stay near Lime Village for a while and then disappear, perhaps to the mountains. In the 1980s another elder reported that sharp-tailed grouse were abundant in the area until fifty years ago when they moved to the Swift River. He added that they occasionally return and speculated that the ashes from a volcano eruption near Anchorage may have caused their departure.

Traditionally Lime Village people have hunted grouse throughout the year but rarely in the spring when grouse are raising young.

Their most active harvest period extends from August through March with August through October being the best time. During those months, grouse are fattest and they tend to feed on gravel in open areas where they are easy to hunt. If grouse are extremely fat during this time, it means a cold winter is approaching.

Lime Villagers hunt grouse early in the morning and in the evening when the birds feed and are easier to catch. An elder explains that grouse eat a small breakfast and lunch, but a large evening meal. Like people, grouse eat a lot in the evening because they don't eat at night. The elder observes that their crop is large at night but small in the morning. Because they feed the most in the evening, it is the best time to hunt them.

Grouse have been taken with bow and arrow, snares, slingshots, and .22 rifles. Snares are placed in trees or on the ground where they are baited with gravel or sand. Spruce grouse have also been killed with rocks thrown by hand. People have walked up to a spruce grouse on the ground and killed it by hitting it on the head with their hand and twisting the bird's neck. An elder comments that spruce grouse are strong, remembering when one knocked him hard enough to cause him to roll backwards several times after he had hit it on the head. Spruce grouse are the easiest grouse to harvest because of their habit of remaining still when danger threatens them. They are said to be especially jittery, however, during windy periods.

After being plucked, grouse are eaten boiled or roasted on a stick. The head and most of the body are consumed, but not the feet because they lack meat. Meatless portions of the wings and the crop are not eaten, but the gizzard has been consumed raw in emergency situations. One way is to remove the rock bag (crop), called *k'geł q'attana,* and simply eat the gizzard raw. An elder remembers eating a raw spruce grouse when he was camping for several days and had no matches to start a cooking fire. He partially froze the bird before completely eating it. He said that the bird tasted "okay." Spruce grouse are good bait for trapping marten.

Grouse feathers have been used for clothing and bedding filling. For a decorative object, the ruffed grouse's crop is blown up to look like a small balloon. Intact grouse tail feathers serve as a fan-like decoration.

CHICKEN (*GALLUS GALLUS*)

gulutsa from the Russian word *kúritsa* "chicken"

The domestic chicken is an introduced bird and not a member of the native avifauna. The term for domestic chicken is known and dates from the nineteenth century when agriculture and chickens were introduced to Cook Inlet by the Russians.

Cranes & Shorebirds

At least some upper Stony River people relate the sandhill crane to the shorebirds because of its long legs and beak, often characteristic of shorebirds. The sandhill crane's body shape resembles that of many sandpipers.

Cranes

SANDHILL CRANE (*GRUS CANADENSIS*)

ndał "they fly," or "plural fly"

Description: Cranes are large birds with long legs, a long neck, and a long, sharp-pointed bill. Adults are generally gray with a bare patch of red skin on the forehead. Some birds are red-brown in color that results from feeding, bathing, and preening in high iron content water. The iron oxide (rust) stain gives the color to the feathers. Juvenile cranes lack the red forehead of the adult. The loud call is distinctive; cranes can often be heard long before they appear. They fly at great heights during migration.

Some inland Dena'ina elders identify two kinds of cranes in the area: a larger, red-crowned bird and a smaller kind lacking the red crown. They describe the smaller bird as "bald-headed" and say that both types of cranes feed and migrate together. The Western scientific system identifies the smaller crane as an immature sandhill crane.

The sandhill crane returns each year to nest in the marshy low-lands of the upper Stony River. Cranes usually arrive in May later than most waterfowl and leave before them in August with the swallows. The cranes are said to circle before leaving to signal the swallows of their departure. The cranes and swallows leave together because the cranes are said to assist the swallows by carrying them under their wings (see Seasonal Cycle).

Traditionally the upper Stony River Dena'ina have harvested sandhill cranes with bow and arrows, snares being the preferred method. Snares are not only easier to use, but are more likely to quickly kill the bird. When using a gun or bow and arrow, people make a special effort to hit the bird in the head because hitting it in the body does not usually kill it quickly and causes the bird to suffer.

Snares are placed at crane feeding grounds, especially where their food is abundant. An elder remembers that his grandmother snared sandhill cranes in the same manner that she snared ducks and other waterfowl.

Hunters would sing a certain song to cranes to which the cranes would sing back and "jump around," allowing the hunter to find their position in order to more easily harvest them. If a crane doesn't fly for a long time, it apparently has a hard time to do so and is easy to approach.

People have been cautioned to be very careful near a wounded sandhill crane or a crane's nest because if the bird's long bill penetrates a person, it may kill him.

The sandhill crane has provided the Lime Village people with food (including eggs) and with feathered-clothing. Water carrying containers, similar to those made from swans and other waterbirds, have been crafted from sandhill crane leg and feet skin.

Shorebirds

Shorebirds migrate to the Lime Village area in the spring, arriving in early May. Some species breed in the area, others are migrants. In part, because some shorebirds are difficult to differentiate, this list of shorebirds found in the area may be incomplete. Especially in times of food shortage, certain shorebirds and their eggs have been harvested for food particularly from beaches. One traditional method is placing snares attached to standing sticks, the length of the stick determining the size of the bird that is caught. Another apparently more commonly used traditional method is tying a trout net between two upright sticks on a beach and driving the birds into the net. Usually only birds larger than the spotted sandpiper (about eight inches in total length) have been kept while the smaller ones are freed unharmed, as the net is closely watched. The small shorebirds would only have been used in a time of extreme food shortage.

Plovers

BLACK-BELLIED PLOVER *(PLUVIALIS SQUATAROLA)*

generally recognized but not recorded

Description: A large heavy-bodied shorebird. In spring identified by the speckled black and white back; black face, chest, and belly; white on the head; and down the neck terminating at the sides of the chest. The undertail is also white. In fall, the body is speckled gray. In all seasons, black-bellied plovers can be identified by the black feathers under the base of the wings. Black-bellied plovers are birds of mud flats and shorelines and would not be common at Lime Village, but the bird has reportedly been seen in the area. It is a very rare visitor in the area.

AMERICAN GOLDEN-PLOVER *(PLUVIALIS DOMINICA)*
PACIFIC GOLDEN-PLOVER *(PLUVIALIS FULVA)*

ggulyit "speed chaser"

Description: Both the American golden-plover and the Pacific golden-plover probably occur in the Stony River area. These two species were separated only recently and we were not able to determine which of these two or both have been identified by the people. The American golden-plover would be expected to be the more common species in the Interior alpine tundra,

AMERICAN GOLDEN-PLOVER

however. Similar in shape to the black-bellied plover, the golden-plover in spring is speckled with a black and golden-yellow back, has a black face, chest, and belly usually including under the tail. The white on the head is over the eye and extends down the neck to the sides of the chest in the American golden-plover and to the tail in the Pacific golden-plover. Fall birds are speckled yellow-brown with Pacific golden-plovers brighter than American golden-plovers. The underwings are uniformly gray in all seasons. Golden-plover migrate to alpine and arctic tundra breeding grounds and may nest near Lime Village. They may be found along the shore, but often in drier grasslands and open tundra. Shorebirds are fast fliers which may have given rise to the Dena'ina name.

The edible golden-plover, that reportedly occurs in wetlands and other environments of the area, has been netted for food but not normally snared. An elder comments that there are traditional stories about this species.

SEMIPALMATED PLOVER *(CHARADRIUS SEMIPALMATUS)*

sank'tnal'ay "the one with a ring around its collar"

Description: Smaller than the other plovers, the semipalmated plover is distinguished by a single dark breast band or collar contrasting with a white belly and throat as the Dena'ina recognize. The back is uniform brown. The bill is yellow-orange with a dark tip and the legs are bright yellow. Semipalmated plovers nest along river, lake, and sea margins usually in open gravel areas where their eggs are hard to find as they closely resemble the gravel. Plovers will pretend they have a broken wing and will hop away from potential predators leading them away from the nest. The semipalmated plover breeds in the Lime Village area.

KILLDEER (*CHARADRIUS VOCIFERUS*)

generally recognized but not recorded

Description: Similar to the smaller semipalmated plover, the killdeer has two dark breast bands and a cinnamon rump and tail. Killdeer are open country birds and may be found on the water's edge or in open fields. They are rare in the Interior and in Southcentral Alaska, but at least one sighting has been reported from the Lime Village area.

Sandpipers

GREATER YELLOWLEGS (*TRINGA MELANOLEUCA*), "TATTLETALE BIRD"
LESSER YELLOWLEGS (*TRINGA FLAVIPES*)

sadya (name refers to its vocal sound)

It has not been determined whether both the greater and lesser yellowlegs occur in the Lime Village area. Probably both species occur there, but the lesser yellowlegs is likely to be more abundant. They may be thought of as one species in the Dena'ina classification system because of their very similar appearance. Yellowlegs breed in the area.

Description: Both species have long yellow legs, a long bill,

GREATER YELLOWLEGS

119

and a gray speckled body plumage. The greater yellowlegs is larger and has a much longer and slightly upturned bill. Lesser yellowlegs are more common in wooded country as along lakes and rivers, while greater yellowlegs are more common in open marsh land and along the sea coast. Greater yellowlegs usually breed in large open areas adjacent to the forest, while lesser yellowlegs breed in small openings within the forest. It is likely that the lesser yellowlegs is the common breeding species in the Stony River area. Greater yellowlegs, while probably occurring there rarely, are more restricted to coastal habitats.

The yellowlegs receives its local English name because it is noisy and tells other birds and wildlife that hunters are in the area. The birds have been snared, netted, and shot for food.

SOLITARY SANDPIPER (*TRINGA SOLITARIA*)

generally recognized but not recorded

Description: Closely related to the yellowlegs, the solitary sandpiper is darker and smaller, but otherwise similar in appearance. It has a dark brown back and heavily streaked neck and breast. There is a white eye ring and its legs are greenish yellow. Solitary sandpipers prefer marshes and woods rather than open beaches and are usually not seen in groups. They do not have the strident call of the yellowlegs. When landing, they often hold their wings upright for a second. This species has been reported for the Lime Village area.

SPOTTED SANDPIPER (ACTITIS MACULARIA)

delvizha "the one that calls *vizh*"

Description: Adults are easily recognized by their solid brown back and white undersides covered with definite spots. The legs and bill are yellow. Juvenile birds lack the spots. Spotted sandpipers fly with bursts of rapid wing beats followed by a short glide as they skim over the water. When on the shore, they frequently teeter or rock back and forth. Spotted sandpipers prefer lake and stream margins. The bird's Dena'ina name refers to one of its characteristic sounds. It breeds in the Lime Village area.

The bird has not normally been harvested for food because of its small size.

UPLAND SANDPIPER (BARTRAMIA LONGICAUDA)

generally recognized but not recorded

Description: A bird of open fields, the upland sandpiper is streaked brown with a relatively small head and long neck, long tail, and long wings. The legs are yellow. A very rare visitor to the Lime Village area with at least one documented sighting.

CRANES & SHOREBIRDS

WHIMBREL *(NUMENIUS PHAEOPUS)*

duzhish dghulggesha "its beak is curved"

Description: A large all-brown sandpiper with a distinctive down-turned bill (Dena'ina name). The head has stripes on the crown and through the eye. During the open water season, it inhabits lake shores and nearby wetlands.

The whimbrel arrives earlier than other shorebirds with the waterfowl. Because of its large size, it has been hunted for food.

BLACK TURNSTONE *(ARENARIA MELANOCEPHALA)*

generally recognized but not recorded

Description: A chunky sandpiper resembling a plover in body shape, the black turnstone is all black above and on the breast, and white on the belly. There are white spots at the base of the bill. In flight, the white wing stripe, back stripe, and base of tail are distinctive. Turnstones are migrants through southcentral Alaska.

In May 1992, one black turnstone was observed along a stream shore near Lime Village. The bird acted strangely and appeared to be lost. Interestingly, in Nondalton, a single black turnstone was seen in the same month.

SURFBIRD (*APHRIZA VIRGATA*)

yudi ghelkala possibly "it has the golden eagle tail"

Description: Similar in body shape to plovers or the black turnstone, the surfbird has an all-speckled plumage with rusty feathers on the shoulders, yellow feet, and bill. In flight, note the white wing stripe and base of tail. Surfbirds nest in alpine tundra and probably breed in the mountains above the Stony River. Interestingly, the Dena'ina name refers to the pattern of white and black on the tail that in fact is similar to that of an immature golden eagle.

Small Sandpipers: *Qenghesh k'ela*

The semipalmated, Western, and least sandpipers are given the same Dena'ina name probably because of their small size and similar appearance.

SEMIPALMATED SANDPIPER (*CALIDRIS PUSILLA*)

qenghesh k'ela "foam tearer" also generic name for small sandpipers

Description: The semipalmated, Western, and least sandpipers are very similar in size and shape. The semipalmated sandpiper has a short rather heavy dark bill and black legs. In the hand, one can see that the toes are partially webbed. The semipalmated sandpiper is a migrant that at least occasionally comes through the Lime Village area.

It is probable that Western sandpipers *(Calidris mauri)* also occur in the Stony River area. They are the commonest small sandpiper nesting on the tundra of Western Alaska and undoubtably come through the Stony River area in migration, probably in larger numbers than the semipalmated sandpiper. However, the appearance of the two species is very similar with minor differences in bill length and plumage coloration that the Dena'ina have recognized them as a single species.

WESTERN SANDPIPER *(CALIDRIS MAURI)*

qenghesh k'ela "foam tearer"

Description: Similar to the semipalmated sandpiper except the bill is longer and appears to droop slightly at the tip. In spring, the scapular or shoulder feathers have a bright red-brown color and there is a wash of the same red-brown on the head. The breast is also more heavily speckled than the semipalmated sandpiper. In fall, they are harder to distinguish except by bill shape. Western sandpipers are the commonest breeding sandpiper on the Yukon-Kuskokwim Delta, west of the Lime Village area, and probably pass through the Stony River on their way to and from their breeding grounds.

LEAST SANDPIPER *(CALIDRIS MINUTILLA)*

qenghesh k'ela "foam tearer"

Description: The smallest sandpiper, the least has streaked brown plumage, a short thin bill, and yellow legs.

The least sandpiper has been sighted in May and may remain to breed during the summer.

SHORT-BILLED DOWITCHER *(LIMNODROMUS GRISEUS)*
LONG-BILLED DOWITCHER *(LIMNODROMUS SCOLOPACEUS)*

kadantsa "tail ?"

Description: Dowitchers are long-billed, stocky-bodied sandpipers that are generally streaked brown above and rusty below with spots and bars on the chest and belly. The names do not correctly indicate the difference between them as the bill measurements of the two species overlap. Short-billed dowitchers nest in Southcentral Alaska while long-billed dowitchers nest in Northwestern and Northcoastal Alaska and, if in the Lime Village area, would be migrants. The two species are best distinguished by the call, a clear "tu tu tu" for the short-billed and a sharp, nasal "keeek" for the long-billed. Dowitchers nest in marshes and often feed in open fields as well as lake and river margins and tidal flats. Long-billed dowitchers are predominant on the Yukon-Kuskokwim Delta and is probably the more common of the two species at Lime Village. However, both species may migrate through the Lime

Village area. Because short-billed dowitchers are more common in Southcoastal Alaska and long-billed dowitchers are more common in Northwestern Alaska, most of the birds moving through the Stony River area are probably the long-billed dowitchers.

WILSON'S SNIPE (*GALLINAGO DELICATA*), "JACK SNIPE"

yuził "sky whistler"

Description: A medium-sized sandpiper with a very long slender bill, stocky body, and short legs. When scared from tall grass, it explodes into the air in a zig-zag flight. In spring, snipe are often heard "winnowing" overhead. This sound, called "whistling," *delził* by the Dena'ina, is created by air rushing through the stiff tail feathers of the snipe as it dives towards the ground.

A traditional story about the Wilson's snipe discourages its use as food. An elder remembers that he and his grandfather ate snipe when they lacked other food. Different versions of the traditional story exist. Elders explain that the bird did something very bad and has a hard time returning to earth. They say that its strange cry is for its lost nest.

Gulls & Gull-Like Birds

Due to similarity in appearance of some adult gulls and plumage similarity and variations in immature gulls, the following Dena'ina gull classification and identification may contain some errors.

Jaegers

LONG-TAILED JAEGER *(STERCORARIUS LONGICAUDUS)*, "SHIT GULL"

nuk'ełvaq'i "the one that induces vomiting"

Description: The long-tailed jaeger is a sleek, gull-like bird with black cap, brown back, white neck and breast, and long central tail feathers. The Dena'ina name refers to the bird's habit of following gulls and harassing them until they regurgitate their prey that the jaeger then snatches in mid air. The bird receives its local English name from its unpleasant habits.

Gulls: *Vach*

refers to the color tan-gray

Gulls are residents of the Lime Village area during the open water season returning in May and leaving before freezeup.

Gull flight-feather shafts have been used for ground squirrel snares and a variety of other snares. Fish nets, fish scoops, and other fishing equipment have been crafted from the shafts. Although shafts from certain other birds have been used, gull shafts appear to have been employed because they are very strong and are more easily obtainable in large numbers than other strong-shafted birds. Special grass blinds have been made at fish camps for harvesting gulls (see Feather Technology). Soft gull feathers have served as bedding filling.

Gull eggs and young gulls have been boiled for food. When cutting fish, people threw scraps to gulls to discourage them from stealing fish and to prevent waste of otherwise unused parts.

GULLS & GULL-LIKE BIRDS

Small Gulls

BONAPARTE'S GULL (*LARUS PHILADELPHIA*)

chilzhena "the one that has a black head"

Description: The smallest gull in the Lime Village area, the adult Bonaparte's has a black head, black bill, gray back, and red legs. Juveniles lack the black head and have a dark spot at the ear and a black band at the end of the tail. They nest in trees and on the ground usually near water.

MEW GULL (*LARUS CANUS*)

shagela vaja "trout's (small fish's) gull"

Description: Generally similar to the larger herring gull, the adult mew gull has a gray back, black wing tips with white spots, yellow feet, and an unmarked, relatively short yellow bill. Juveniles are brown with all dark wings and a dark tail. Second-year birds are grayer with black wings and a dark bar at the end of the tail. They nest in trees and on the ground in marshes. The bird's Dena'ina name refers to its habit of eating small fish.

Large Gulls: *Vach kegh* "big gull"
(generic name for the large gulls)

HERRING GULL (*LARUS ARGENTATUS*)

łiq'a vach, łiq'a vaja "salmon gull" (probably the specific name for the herring gull, the most common gull of the area)

Description: Adult is a large gull with silver-gray back, black wing tips with white spots, yellow bill with a red dot on the lower mandible, and pink legs. Juveniles are dark brown or gray-brown with black wings and solid dark tail. Second- to fourth-year birds are grayer and streaked, the bill gradually becoming lighter with age. The tail is initially dark, but becomes lighter with a dark terminal band by the third year, and all white by the fourth year. Herring gulls are common inland along rivers and lakes.

GLAUCOUS-WINGED GULL (*LARUS GLAUCESCENS*)

vach kegh "large gull"

Description: Adult resembles similar sized herring gull, but the wing tips are pale gray with white spots and the bill is heavier. Juveniles and immatures are similar to but lighter than comparable aged herring gulls. Glaucous-winged gulls are more common along the coast, but occur inland on some river systems.

Glaucous Gull *(Larus hyperboreus)*

vach kegh "large gull"

Description: The adult is an all-white gull with a very pale-gray mantle, yellow eye, large yellow bill, pink legs, and white primary wing feathers. Immature birds in the first through fourth years are often duskier, but all have white wing tips. Glaucous gulls are primarily coastal birds but often venture up river systems and are probably not uncommon on the Stony River.

Terns

ARCTIC TERN (*STERNA PARADISAEA*)

ch'ink'nul'ay "it keeps its head erect," also called *shtlaq* "?"

Description: A long streamlined gray and white bird with a black cap, red bill and legs, and a long forked tail. The rowing flight with slow wing beats is distinctive. Terns usually nest in colonies on islands in marshes or on isolated sandy islands, the water protecting their nests from land predators. They will dive bomb intruders.

The arctic tern, also a fish eating bird, arrives later and leaves earlier than the gulls. Tern eggs have been collected for food.

Climbing Birds

Woodpeckers:
Kuntsulya "stomach ?"

Although four species of woodpecker are recognized by the Lime Villagers as permanent residents of the Stony River area, their names do not correspond directly to each species recognized by Western scientific classification. Any woodpecker is named *kuntsulya* "stomach ?"; the "alder woodpecker" *qenq'eya kuntsulya;* and the "willow woodpecker" *ch'etl' kuntsulya* may be the downy and hairy woodpeckers, while the term "spruce woodpecker" *ch'vala kuntsulya* refers to the three-toed and black-backed woodpeckers.

Northern Flicker (*Colaptes auratus*)

tsenel "nose-wedge"

Description: A large woodpecker that often feeds on the ground. The back has distinct dark bars and the undersides are spot-

ted; there is a black bar across the upper chest. Males have a black moustache mark extending from the base of the bill. The underwing and undertail linings are bright yellow. Flickers nest in hollow trees. They are becoming more common in Interior

Alaska and may be found in the Lime Village area. The name *tseneł* is used for flicker in Upper Cook Inlet Dena'ina; this is a widely distributed name in the Athabascan language family.

DOWNY WOODPECKER (*PICOIDES PUBESCENS*)

ch'etl' kuntsulya "shrub or willow woodpecker"

Description: The smallest woodpecker, patterned black and white with a relatively short bill. There are a few black bars on the outer white tail feathers. The male has a red patch on the back of the head. Usually inhabits deciduous forests, but occurs in mixed woods as well. Often feeds on insects in willow and alder brush and woody plant stems.

HAIRY WOODPECKER (*PICOIDES VILLOSUS*)

qenq'eya kuntsulya "alder woodpecker"
eseni kuntsulya "cottonwood woodpecker"

Description: Similar in appearance to the downy woodpecker with its black and white patterned plumage, the hairy is larger, has a proportionately larger bill, and lacks bars on the outer white tail feathers. Males have a red patch on the back of the head. Usually inhabits deciduous or mixed woodlands. Feeds on insects in birch and cottonwood trees, alder, and large willow bushes.

THREE-TOED WOODPECKER *(PICOIDES TRIDACTYLIS)*

ch'vala kuntsulya "spruce woodpecker"

Description: A large dark woodpecker with fine dark barring on the white flanks and on the back. Males have a yellow crown. Feeds on insects in spruce forests.

BLACK-BACKED WOODPECKER *(PICOIDES ARCTICUS)*

ch'vala kuntsulya "spruce woodpecker"

Description: Like the three-toed woodpecker, except the back is all black. Both species have only three toes on each foot. Males have a yellow crown. Feeds on insects in spruce forests.

After being dried, a woodpecker's skin has been attached to a child's clothes as an amulet to help the child (usually a boy) become a good wood chopper. The yellow-headed (three-toed and black-backed) woodpeckers appear to be preferred as amulets over the red-headed (downy and hairy) woodpeckers.

Versions of a traditional story explaining how woodpeckers learned the right food to eat are found in Wassillie (1980, 31–33) and Tenenbaum and McGary (1984, 75) in an account told by Mary V. Trefon.

Creeper

Brown Creeper *(Certhia familiaris)*

kuntsulya gguya "small woodpecker"

The upper Stony River Dena'ina consider the brown creeper to be a kind of *kuntsulya*. The creeper and nuthatch are called by the same term.

Description: A small brown and white patterned bird with a long down-curved bill. The tail is stiff like a woodpecker's and the bird climbs trees from the bottom upwards searching for insects in crevices in bark. It is a permanent resident occurring in low numbers in the area. An elder reports having seen the brown creeper infrequently during his life time.

Nuthatch

Red-breasted Nuthatch *(Sitta canadensis)*

kuntsulya gguya "small woodpecker"

Description: A small gray-blue climbing bird with a black cap and black line through the eye. Adults have a rusty breast and belly, the male being darker than the female. Nuthatches generally climb down the tree rather than up, but often hunt for insects along branches as well. It is a winter resident that occurs in low numbers in the Lime Village area.

Possibly because they may be uncommon, no uses have been reported for the brown creeper and red-breasted nuthatch.

Scavengers

The scavenger birds—gray jays, magpies, and ravens—are grouped together by the Dena'ina because they commonly steal food and cause trouble. They also occur in traditional stories together.

Gray Jay (*Perisoreus canadensis*), "Camp robber"

ch'k'naqałch'eya "?" suggested meaning
"one that eats a little"

Description: Gray jays are fluffy gray birds common around human settlements. The forehead and undersides are white and the back is dusky gray. Juveniles are all dark gray. These permanent residents of the Lime Village area are brazen as they approach humans and steal food from them, often by taking small pieces away and caching them for later consumption.

As is true of almost all birds not normally hunted for food, the gray jay can provide emergency food. An elder remembers a time in her youth when gray jays provided urgently needed food. Another elder recalls from his youth the old people saying that they often ate gray jays when other food was lacking. They boiled or roasted them on a stick, but they did not singe them first like waterfowl. Not only have gray jays provided emergency food, they tell a hunter by making the sound *tsa tsa* (wait, wait) that he will have hunting success.

An elder observes that the gray jay likes to tease the saw-whet owl and that the owl does not respond. He speculates that the gray jay teases the owl because it likes the owl.

The gray jay is notorious for stealing food and has often been the first game of children learning to shoot. An elder comments that the meaning of the Dena'ina name for the gray jay refers to the fact that the bird makes a mark on a person's meat by eating part of it.

See Tenenbaum and McGary (1984, 87, 125) for references to the gray jay. The first story called "Raven and the Bird People" is told by Gulia Delkittie. The second story, "Raven," is by Antone Evan. In these stories, the gray jay is closely associated with the raven and magpie.

BLACK-BILLED MAGPIE *(PICA HUDSONIA)*

q'ahtghal'ay "the proud one"

Description: An unmistakable large bird with long tail and glossy black and white plumage. The wings and tail often shine with blue or green iridescence. A permanent resident of the Lime Village area.

As is true of common ravens, Lime Villagers have tamed magpies as pets and hunting assistants (see Taming and Training Birds). A tamed magpie may bring a hunter good luck by helping him sight game. People have watched the flight of wild magpies to find game. See Tenenbaum and McGary (1984, 87, 145) for two traditional stories involving magpies.

COMMON RAVEN *(CORVUS CORAX)*

chulyin "the one that eats shit"

Description: An all black, very large bird with a heavy bill and shaggy throat. Their many voices are distinctive from raucous caws to

bell-like calls. In spring they often perform aerial acrobatics using updrafts from cliffs. Ravens are permanent residents of the Lime Village area usually nesting along cliff faces and spending the winter often near human habitation where food is more available.

Although not edible except in an emergency, the bird's feathered-skin has been valued for making clothing and the feathers for headdresses (see Bird Clothing). Reportedly if a young girl wears a raven-feather hat, the girl's hair will not turn gray at a later age. Raven-feather shafts have been used in the construction of ground squirrel snares and other snares used on land.

SCAVENGERS

The bodies of ravens have been left near harvested game such as waterfowl and fish to frighten scavengers. However, to scare away ravens, sharpened sticks, axes, and knives have been left near harvested meat. A traditional belief is that if a person kills a raven without using it, he will have bad luck.

When trained as a young bird, a raven helps its trainer to hunt. It circles around game to show the hunter where it is located. A tamed raven can talk to a person in the person's language and give the person needed information. A wild raven can signal game in an area but not as efficiently as a tamed raven that also warns people of danger. A trained raven might travel with another person it knows but always returns to its trainer if he has treated him well.

Although infamous for stealing food, ravens have been said to be smarter at obtaining their own food than hawks and owls.

Certain actions of wild ravens are signs to people. For example, if a raven flies in a circle, somebody will have good hunting luck. When a certain raven sound, *ggagga*, is heard by a hunter, this means that he will obtain game. (The word *ggagga* means "creature, four-legged animal" and "brown bear" in the Inland Dena'ina dialect, J. Kari 1977, 23.) A hunter who sees a raven roll over in the sky will experience good hunting luck. If a raven "hollers" at night, misfortune will occur. A person who imitates a raven sound will be poor and have ragged clothes like the raven. Three of these five traditional beliefs refer to the raven's sounds, indicating their importance.

Culturally, people both respect and disdain the raven. While admired for his intelligence and ingenuity, he is disliked among other things, for his thievery, trickery, and bad manners. Many traditional stories exist about the raven. When animals were people, he is said to have been a great shaman who both helped and harmed other people. The following description is from *Raven Returns the Sun* (Wassillie 1980, 28). "At the time of this story, Raven was everything. He was the creator. He was the gentleman, the thief, the crook, and also the murderer." See Wassillie 1980 and Tenenbaum and McGary 1984 for other traditional stories about the raven.

Small Birds

SMALL BIRDS

Winter Birds:
Hey ch'ggaggashla

Chickadees: *Ch'ggagga* "creature"

BLACK-CAPPED CHICKADEE *(POECILE ATRICAPILLA)*

tsidut'aq'a "bare headed"

Description: A small gray, black, and white bird, with black cap and throat, gray wings and tail with white feather edgings, and white below. Chickadees are tame and can easily be lured to take food from the hand. Their familiar call of "chick-a-dee-dee-dee" gives them their English name. The spring song is a series of three clear whistles. Prefers deciduous and mixed forests. A permanent resident of the Lime Village area.

BOREAL CHICKADEE (*POECILE HUDSONICA*)

ch'ggagga "creature"

Description: Similar in size
and habit to the black-
capped chickadee, the bo-
real has a brown cap, less
black on the throat, a
gray-brown back, rusty
sides, and pale belly. Its call
is buzzy sounding compared
to the black-cap and it does not sing
a whistled spring song. Prefers coniferous forests. It is also a
permanent resident.

Although chickadees apparently have no material value except as
emergency food, more meaningful actions and beliefs appear to be
associated with the chickadee than with other small birds. Following are some examples: If a chickadee hits a window or enters a
house, the household can expect a visitor or, if no visitor, good hunting luck. If neither happens misfortune may occur. Feathers left on
the house increases the significance of the events. Although this
sequence of events is especially meaningful when a chickadee par-
ticipates, other small birds may initiate the events in the same
manner.

If a person hears a chickadee when he is in the forest away from
home, the person is being thought of by someone or he will receive
a visitor.

A black-capped chickadee's spring song that occurs when the days
turn long is said to be its cry for winter. Reportedly it is unlucky to
find a chickadee's nest.

The literal meaning of the chickadee's name *ch'ggagga* according
to some people also indicates the importance of the bird because
the Inland Dena'ina name for the lifeform category "bird" is
ch'ggaggashla or "little creature." An elder speculates that the
chickadee's Dena'ina name refers to "little bird."

Chickadees are prominent in traditional Dena'ina stories. Like the raven, gray jay, and magpie, with which it is sometimes associated in stories, the chickadee can be troublesome or helpful. Stories involving chickadees occur in Tenenbaum and McGary (1984, 13, 25, 87). The story "Two Women" by Antone Evan explains how the chickadee received its cap. The black-capped chickadee's Dena'ina name which means "bare-headed" may refer to this event.

People feed chickadees by hanging small pieces of animal fat near their homes.

Grosbeak

PINE GROSBEAK (*PINICOLA ENUCLEATOR*)

qidudya (name refers to its vocal sound)

Description: A robin-size fluffy appearing bird with dark wings and tail and white wing bars. The bill is short and thick. The adult male has a pink or rosy-red head, breast, and back. Females and juvenile males range from orange to yellow to greenish color on the head, the rest of the body being gray. Pine grosbeaks breed and feed in spruce forest but also eat the buds and fruits of deciduous shrubs and trees. The pine grosbeak is a permanent resident of the upper Stony River area.

A Lime Village resident relates that he sees pine grosbeaks feeding along streams. According to an elder, the bird's whistled, cheerful, three-note call may be the reason for its name.

Summer Birds:
Shan ch'ggaggashla

Kingfisher

BELTED KINGFISHER *(CERYLE ALCYON)*

chik'dghesh "ragged head"

Description: Both adults are slaty blue-gray above and white below with a blue-gray chest band. The female has a rusty band across the belly. Kingfishers have proportionately large heads with a ragged crest, thus the Dena'ina name. Kingfishers perch over water or often hover before they dive for small fish. Their rattling call is distinctive. They nest in burrows dug into sand banks.

The kingfisher is a summer resident of the upper Stony River area where it fishes along streams and in lakes. It is not used for food due to a traditional story in which the kingfish teamed up with the pika to rescue many missing animals. In the story by Alexie Evan (Tenenbaum and McGary 1984, 179–190) the kingfisher is called *Ch'iduchuq'a*. Note however in Peter Kalifornsky's version of a similar story (Kalifornsky 1991, 72–77) the bird is the northern shrike.

Flycatchers: *Dehvava* "their own dry fish"

Flycatchers perch on an exposed branch and fly out after insects, returning usually to the same perch. They are often best recognized by their calls than by their olive-drab plumage.

OLIVE-SIDED FLYCATCHER *(CONTOPUS COOPERI)*

vava nihi "the one that says dry fish" or *dehvava* "their own dry fish"

Description: A large olive-gray flycatcher with a proportionately large head, light below, with a dusky vest across the breast, and white flank feathers often visible above the folded wings. The call is distinctive, as the Dena'ina recognize, and is interpreted by English speaking people as "hip three cheers."

ALDER FLYCATCHER *(EMPIDONAX ALNORUM)*

generally recognized but not recorded

Description: A small gray-green flycatcher with a lighter slightly cream-yellow belly, two light wing bars, and a distinct white eye ring. Its call is a described as "wee-bee-o," falling abruptly at the end.

The olive-sided flycatcher is the more culturally-significant flycatcher because its call *vava*, a Dena'ina word meaning "dried fish," signals that the salmon are traveling up the Stony River. It also calls when the suckers begin to run and stops when they are abundant. Although the call of the olive-sided flycatcher consists of three syllables, at a distance only the latter two can easily be heard. These two loud whistles may have given rise to the Dena'ina name *vava nihi*.

At least some Stony River people as well as Nondalton people identify the golden-crowned sparrow as the bird or one of the birds that foretells the fish run.

Swallows: *Kałja* "tail ?"

Swallows are small swift flying birds that spend most of their time on the wing hunting flying insects. They nest in cavities that they either find in hollow trees or create by digging holes in sand banks or making mud nests. The term for violet-green swallow serves as the term for the class of swallows.

TREE SWALLOW (*TACHYCINETA BICOLOR*), "TOM SWALLOW," "BLUEBACK"

tl'ałghak'a "firedrill ?" or "flint ?"

Description: Dark above and white below, adults are glossy blue or blue-green on the head down to below the eyes and on the back. Fall birds are greener and juveniles are browner with a dusky wash across the breast. They nest in hollow trees, boxes, or in other cavities found around human settlements.

The name appears to contain the root for "firedrill" *tl'ał*.

SMALL BIRDS

VIOLET-GREEN SWALLOW *(TACHYCINETA THALASSINA)*

kaḷja "tail ?"

Description: Similar to the tree swallow, dark above and white below, but the white of the face extends behind and above the eye; the head of the adult is shiny purple and the back is glossy green. The white of the flank extends upwards onto the base of the tail making two white patches on the rump visible in flight. Juveniles are browner above and clear below. They nest in similar sites as the tree swallow. A cliff above Qeghnilen on the Stony River, Kaḷjana, is named for this swallow.

BANK SWALLOW *(RIPARIA RIPARIA)*

vest'ugh kaḷja "beneath the bank swallow"

Description: All dark brown above and white below with a definitive dark brown band across the breast. Bank swallows excavate tunnels in vertical dirt banks and nest in colonies of several to hundreds in the same bank.

CLIFF SWALLOW *(HIRUNDO PYRRHONOTA)*

qenk'dengheḷts'etl'i kaḷja "swallow that daubs something (mud)," also possibly *tsanenh kaḷja* "cliff swallow"

Description: Dark blue on the crown and back, pale below, this swallow has a buffy rump, rusty face and chest, and a white forehead. They build nest from globs of mud (as the Dena'ina name indicates) that are anchored underneath an overhang often on a man-made structure but originally on cliff faces.

Swallows are migrants to the Lime Village area with the tree and violet-green swallow arriving in early May and the bank and cliff swallow often not arriving until the end of May or first of June. They do not remain in the area long after the young are fledged and are usually gone by the middle of August.

Swallows are thought of as very good birds because they, more than most other birds, help people by eating annoying insects near human dwellings. The upper Stony River Dena'ina have continued the tradition of making nesting boxes for swallows and placing them close to or on their own houses.

Larks

HORNED LARK (*EREMOPHILA ALPESTRIS*)

generally recognized but not recorded

Description: Horned larks are birds of open fields, beaches, and alpine meadows. They have a pale-yellow face, black mask through the eye, black tufts on the head, and a black breast band. The back is uniform pinkish-brown and the tail dark with white outer feathers. They migrate in loose groups often in association with longspurs and pipits and have been seen in the Lime Village area.

Dipper

American Dipper (*Cinclus mexicanus*)

tatsilqit'a "one who has its head in water"

Description: A very stubby heavy-bodied all dark gray bird with a short tail and wings. Dippers are permanent residents of moderate to fast flowing streams along cliffs and other rocky places in the upper Stony River area. They move completely submerged along the bottom of streams where they hunt for insect food.

Although Lime Villagers have not normally harvested American dippers for food, they have, however, made waterproof containers from its leg and foot skin.

American dippers are said by at least some Lime Village Dena'ina to be related to rusty blackbirds and common ravens because of their similar, dark color. In *Raven and the Bird People* told by Gulia Delkittie, the American dipper, common raven, magpie, rusty blackbird, and gray jay are associated while in *Raven* told by Antone Evan, the above mentioned birds except the rusty blackbird appear together (Tenenbaum and McGary 1984, 25, 87).

An upper Stony River Dena'ina reports that in a traditional story, the American dipper is wounded when wind blows rocks from a cliff. There is also a song about the bird.

Thrushes: *Shinggach'* or "Big-eyed Birds"

The Swainson's, gray-cheeked, and hermit thrushes are all super-ficially very similar. All have dark spots on a light breast, an olive-brown head, back, and wings; the underwings have a pale orange stripe visible in flight. These three species often feed during twi-light hours and their relatively large eyes may assist in seeing in dim light conditions. All have pleasing songs. The varied thrush and American robin are larger and have reddish or orange breasts, and the wheatear is a Eurasian migrant that spends its summers on alpine tundra in Alaska. All are summer residents of the upper Stony River area. The Dena'ina thrush names, except possibly the Swainson's thrush and the American robin, probably refer to the bird's song or call.

The varied thrush arrives in the local area as bare ground first appears, followed by the robin and finally the hermit, Swainson's, and gray-cheeked thrush that come after the river ice above Lime Village has moved below the community.

Lime Villagers have partially tamed thrushes by coaxing them with food placed close to and eventually inside their homes. A woman remembers keeping a robin and gray-cheeked thrush during dif-ferent winters in her house. The hermit thrush and at least some of the other thrushes help people by eating insects and insect eggs from drying fish.

GRAY-CHEEKED THRUSH (*CATHARUS MINIMUS*)

ggezhaq name refers to its vocal sound

Description: Similar to the Swainson's thrush, the gray-cheek lacks the eye ring and buffy coloration on the face and breast. Instead, the spots are on a whiter or grayish-white background. The tail is the same color as the back.

The song is a descending series of wheezy trills that usually rises at the end.

Reportedly the gray-cheeked thrush calls *"ggezhaq"* upriver from Lime Village when the river ice has left there. When all the ice has left the river, it calls a different way.

SWAINSON'S THRUSH *(CATHARUS USTULATUS)*, "BIG-EYED BIRD"

shinggach' possibly "it's staring at me"

Description: Swainson's thrush is distinguished by the presence of a buffy eye ring and a buffy background to the spotted breast. The tail is the same color as the back. The song is a rolling, ascending series of trills.

The name seems to mean "it is staring at me," and is the source of the local name Lime people call it, "big-eyed bird."

HERMIT THRUSH *(CATHARUS GUTTATUS)*

kentuchila "meadow-water-moist" (perhaps refers to its call or song)

Description: Similar to Swainson's thrush, the hermit thrush has a reddish-brown tail and a generally warmer brown plumage than the other similar thrushes. Its song is a series of ever changing flute-like whistles and complicated trills.

AMERICAN ROBIN (*TURDUS MIGRATORIUS*)

kałnay "-?-"

Description: A large thrush with gray head, back, wings, and tail. The breast is orange-red and the belly is white. There are white tips to the outer tail feathers. Young robins have spotted breasts showing their relationship to the other thrushes. The song is a rolling series of loud rising and falling disconnected phrases.

An elder observes that robins nest both on dry ground under trees and in trees, while the other thrushes nest only in trees. Another person relates that a robin will return to a nest with eggs touched by humans and continue to care for the eggs and young birds.

The robin's noisy call, interpreted as "tut, tut" with variations, is the subject of a traditional Dena'ina story. Briefly, a woman with a son had a husband who left for days at a time. The robin learned that the husband was having an affair with another woman and made the call to tell her. While the event happened during the time when birds and humans spoke the same language, the call has been passed down from robin to robin to this day. Reportedly some people have been able to understand the robin since early time.

VARIED THRUSH (*IXOREUS NAEVIUS*)

deltl'ishi "the one that calls *tl'ish*"

Description: A large thrush with an orange breast and belly, orange wing bars, and a bar across the breast. Males are gray above with a black mask and black bar across

157

the breast; females are duller brown with their dark markings much subdued. The song consists of elongated buzzy or trilled notes each on a single pitch. Each phrase sounds like the bird is uttering several frequencies at once, but it actually is trilling from low to high notes in very rapid succession. Trills are given on a series of different pitches and sound like an old fashioned doorbell ring.

Thrush-Like Birds

Because of certain thrush-like characteristics and habits, at least some Dena'ina relate the northern waterthrush and the fox sparrow with the thrushes. Not only do both species look similar to the Swainson's, gray-cheeked, and hermit thrushes but like the thrushes, they usually feed on the ground. They also eat insects and insect eggs on drying fish.

NORTHERN WATERTHRUSH (*SEIURUS NOVEBORACENSIS*)

nent'ughazh "underground round object"

Description: A warbler that looks more like a thrush with a solid brown back and white underparts streaked with dark brown. There is a distinct white line over the eye. Found on or near the ground, usually near flowing water, the waterthrush bobs its tail continuously as it hunts for insects.

FOX SPARROW (*PASSERELLA ILIACA*)

dushidaghiłchiq'i "why does it scold me?"

Description: A large heavily streaked sparrow with a reddish-brown tail. Differs from the thrushes by its large heavy cone-shaped bill as opposed to the more slender bill of the thrushes.

One or more of the bird's relatively loud songs or calls may be the reason for its Dena'ina name.

An elder remembers that a fox sparrow that accidentally missed the fall migration was kept in a person's house throughout the winter. It went outside periodically, but returned to eat the crackers, bread, and rice that it was fed. The elder believes that the bird would have frozen if it had not been taken into the house and cared for. Its name, *dushidaghiłchiq'i,* is charming and unusual in that it is phrased as a question.

NORTHERN WHEATEAR *(OENANTHE OENANTHE)*

generally recognized but not recorded

Description: A small bird with a white rump and base of tail and a dark bar at the end of the tail. Males are gray above and white below with a black mask and black wings. Females are brownish gray above and pale below. Wheatears nest in alpine tundra where they appear nervous as they fly about and flick their tails. The wheatear has been seen above Lime Village.

SMALL BIRDS

Brightly Colored Small Birds

At least six warblers (including the northern waterthrush) have been reported to summer in the upper Stony River area. As a group, the Dena'ina do not appear to classify them together other than as *shan ch'ggaggashla* "small summer birds," which includes sparrows and other small summer breeding birds of the area. However, warblers that are obviously at least partially yellow are grouped under the Dena'ina name, *k'eltseghay, ltseghay* "the yellow one." The category includes the yellow, Wilson's, yellow-rumped, and possibly the orange-crowned and arctic warblers that have less noticeable yellow feathers. The yellow birds are especially enjoyed for their bright colors combined with their "happy" songs, the latter being a characteristic of many summer birds. The warblers, some of the last returning migrants, are a sign that summer has or is about to arrive.

No use other than emergency food is reported for these warblers and the ruby-crowned kinglet.

Kinglets

RUBY-CROWNED KINGLET (*REGULUS CALENDULA*)

deghgich "hanging cartilage or hanging rustling noise" or *jatnuldeli* "one with a stripe of red"

Description: A tiny gray-green bird, pale below, with white wing bars and a conspicuous white eye ring. Only the male has the ruby-red crown feathers that are usually concealed unless the bird is excited. The song is unexpectedly loud for such a small bird and starts with several high notes, then several lower notes, followed by a rolling warble. The ruby-crowned kinglet is migratory, often arriving in early May, sooner than many small summer birds.

Old World Warblers

ARCTIC WARBLER (*PHYLLOSCOPUS BOREALIS*)

k'eltseghay, ltseghay "the yellow one"

The Dena'ina name for the yellow-colored warblers is generic; the yellow warbler is probably the type and has the generic name as its specific name.

Description: The arctic warbler migrates from Eurasia each spring to nest in Alaska and returns in the fall. It is related to the warblers of Europe and Asia and not to the following North American warblers according to the Western system of classification. The arctic warbler is olive gray-green above and pale yellowish below. There is a distinct dark line through the eye and a pale line over the eye. The song is a series of buzzy notes. Inhabits willow and alder brush often at higher elevations.

New World Warblers

ORANGE-CROWNED WARBLER (*VERMIVORA CELATA*)

k'eltseghay, ltseghay "the yellow one"

Description: A small almost uniformly olive gray-green bird without any conspicuous markings. Some

161

individuals are very yellow-green in color but those in Lime Village area are gray-green. Usually there is a light line over the eye. In adults, the breast is streaked with dusky olive gray. Most males and some females have a patch of concealed orange feathers on the crown that are visible only when the bird raises its crown feathers in excitement.

Yellow Warbler (*Dendroica petechia*)

k'ełtseghay, ltseghay "the yellow one"

Description: Olive-yellow above and bright yellow below, yellow wing bars, and yellow spots on the tail feathers distinguish this warbler from all others. The eye appears large with a bright yellow eye ring. Adult males have fine reddish streaks on the breast.

Yellow-rumped Warbler (*Dendroica coronata coronata*)
Myrtle Warbler

k'densuya "the one that makes a rattling noise"

Description: The bright yellow feathers at the base of the tail identify this warbler. Adults have yellow on the crown and below the shoulder, but most of the plumage is black, gray, and white. Males have a bluish-gray back, black mask and breast streaks. Females are duller gray and often brownish-gray on the back and have a duller mask and breast streaks.

BLACKPOLL WARBLER (*DENDROICA STRIATA*)

generally recognized but not recorded

FEMALE

Description: Males are streaked black and white and have a black cap. The legs are orange. Females and juveniles are duller, browner, and tinged with yellow. They lack the bright yellow feathers of many other warblers. Blackpoll warblers nest throughout forested parts of Alaska. Their song is at a very high frequency and often hard to hear.

WILSON'S WARBLER (*WILSONIA PUSILLA*)

k'eltseghay, ltseghay "the yellow one"

Description: A small warbler, bright yellow below and olive green above. Males have a shiny black cap. Distinguished from the yellow warbler by the lack of any yellow wing bars or tail patches, and from the orange-crowned warbler by lack of a line over the eye, bright yellow breast and belly, and no orange feathers on the head.

Sparrows & Buntings

Sparrows are brown, usually heavily streaked birds with cone-shaped bills capable of crushing small seeds. Most nest on or near the ground. Sparrows are spring migrants to the upper Stony River area where they usually arrive in May. Except as emergency food, no material use is recorded for them.

SMALL BIRDS

AMERICAN TREE SPARROW (*SPIZELLA ARBOREA*)

ch'nutch'ish (refers to a rustling or scraping noise)

Description: The tree sparrow has a clear unstreaked gray breast with a single central spot. The crown is reddish brown. Tree sparrows nest on mountain sides just below or at tree line and often winter in the valleys.

SAVANNAH SPARROW (*PASSERCULUS SANDWICHENSIS*)

ch'ich' qunsha "ground squirrel that goes *ch'ich'* (scraping noise)"

Description: Streaked above and below, the savannah sparrow is distinguished by a light yellow line over the eye. It is a bird of open habitats and grasslands. Its Dena'ina name may refer to the ground squirrel because people have heard it in the mountains when harvesting ground squirrels.

LINCOLN SPARROW (*MELOSPIZA LINCOLNII*)

tl'ujech'a "tail that dances"

Description: Streaked brown above and pale below, the Lincoln's sparrow has a gray wash to the face and the buffy breast is streaked with very fine dark streaks, unlike the coarser streaks of the savannah and fox sparrows. The Dena'ina name may refer to a story in which the Lincoln's sparrow pumps its tail up and down in flight.

GOLDEN-CROWNED SPARROW *(ZONOTRICHIA ATRICAPILLA)*

tsik'ezdlagh (name refers to its vocal sound)

Description: A large sparrow streaked brown above and clear brownish-gray below. Adults have a black cap with a yellowish streak in the center of the crown. Juveniles have a brown cap and very faint yellow streak. Their three-noted song, interpreted in English as "oh dear me" is distinctive.

Lime Villagers have seen the golden-crowned sparrow especially in the mountains. The bird has also been observed near Lime Village.

Reportedly, if a hunter hears a certain call of the golden-crowned sparrow meaning "animal liver" (not the three noted song), it is a prediction that the hunter will catch an animal. The bird is said to come to a person who harvests an animal. If a person tells the golden-crowned sparrow his name, the bird will attempt to imitate it, but cannot really talk to a person like an owl. The bird is one of two species whose song tells people that the salmon are coming; the olive-sided flycatcher is the other.

WHITE-CROWNED SPARROW *(ZONOTRICHIA LEUCOPHRYS)*

ggulush didi "?"

The name has been said to refer to the black and white stripes on the head; however *ggulush* is also term referring to "Tlingit people."

Description: A large sparrow, streaked brown above and clear gray below. Adults have a definite black-and-white striped head; juveniles have brown instead of black head stripes.

A traditional belief is that if a hunter hears a white-crowned sparrow change his song, it means that the person will have hunting success. An elder relates that he has had that experience. Another person observes that the white-crowned sparrow makes a noise only once at night and then waits until morning to begin singing. White-crowned sparrows assist people by eating maggots and insects on drying fish. People have tamed white-crowned sparrows.

DARK-EYED JUNCO (*JUNCO HYEMALIS HYEMALIS*)
SLATE-COLORED JUNCO

k'delt'edi "the one that call *t'et* (smacking sound)"

Description: Adult juncos are slate gray above and white below with conspicuous white outer tail feathers. Males are darker with an almost black head and neck; females are grayer and sometimes brownish color. Juveniles are streaked above and below showing their relationship to the sparrows.

Juncos come earlier than most small summer residents and leave later. An elder relates that the bird has another Dena'ina name that refers to "half winter," indicating its longer time in the area than many other migrants.

LAPLAND LONGSPUR (*CALCARIUS LAPPONICUS*)

generally recognized but not recorded

Description: Longspurs are streaked like a sparrow, but the hind toe nail or claw is very long. The outer tail feathers are white and the bill is pale yellow in spring. Males in spring have a black crown, face, and breast, a white line from the eye down

the neck, and a chestnut red nape of the neck. Females are duller but have a hint of rusty red on the nape. In migration longspurs move in flocks along open country, shorelines, beaches, and grasslands. They breed in alpine and arctic tundra.

The Lapland longspur is an early spring migrant that passes through the Lime Village area. It arrives between mid April and early May when snow still partially covers the ground and before most other song birds arrive. People report having seen them in the spring waterfowl feeding areas and in the community. The Dena'ina name is undocumented or may be forgotten.

SNOW BUNTING (*PLECTROPHENAX NIVALIS*), "SNOW BIRD"

nehggux "grayish"

Description: Males are black and white; females gray to brown and white. In flight, the black wing tips and central tail feathers contrast with the white base of the wings, outer tail feathers, and base of tail. In migration, birds become browner on the head and back. Snow buntings are birds of open country and migrate along shores and beaches. They breed in arctic and alpine tundra.

SMALL BIRDS

Snow buntings migrate through the upper Stony River area at the end of March or in April. When food was in short supply, people have hunted the small birds with bows and arrows or more recently guns, being careful to aim at the head to preserve the meat. Another method for catching snow buntings was to prop up a baited container such as a birchbark basket or pan with a stick. When the bird entered the pan, the stick was released to trap the bird. Snow buntings have normally been stewed.

Blackbirds

RED-WINGED BLACKBIRD (*AGELAIUS PHOENICEUS*)

generally recognized but not recorded

Description: Males are all black medium-sized birds with brilliant red patches at the bend of the wing or shoulder bordered by yellow. Females are heavily streaked with brown and lack the red wing patches but may be distinguished from sparrows or finches by the long tapered sharp bill. First year male birds are brown and streaked but have some red on the shoulder.

Several people have reported sighting red-winged blackbirds in the Lime Village area but they are not regular migrants.

RUSTY BLACKBIRD *(EUPHAGUS CAROLINUS)*

ch'qełch'ah (name refers to its vocal sound)

Description: A medium-sized all black bird identifies the breeding male. Females are all gray. The plumage of fall birds and juveniles has rusty brown feather edges giving the bird its English name. The eye is yellow.

The rusty blackbird, an inhabitant of wetlands, is a spring migrant to the Lime Village area. Its Dena'ina name imitates the bird's song.

The rusty blackbird is mentioned briefly in the traditional story *Raven and the Bird People* told by Gulia Delkittie (Tenenbaum and McGary 1984, 87). Interestingly, it is associated with the raven, magpie, gray jay, and American dipper that are also dark-colored birds.

Nomadics

Except for the Bohemian waxwing, the following group of birds is composed of finches whose abundance in any one area is more or less dependent on food supply. Most northern tree species do not produce abundant seed crops every year in any one location. Nomadic finches move from place to place to take advantage of the current seed crop and often breed in those same areas.

Waxwing

BOHEMIAN WAXWING (*BOMBYCILLA GARRULUS*)

k'eghunjech'a "war dancer"

Description: A medium-sized gray bird with a conspicuous crest. Waxwings have a black mask and throat, black wings and tail, and chestnut red under the tail. There is a yellow band at the tip of the tail and a white wing bar. The English name is derived from the red "wax" at the tip of some inner wing feathers. Waxwings are nomadic birds that follow the ripening of berries and other foods they prefer.

Lime Villagers say that the Bohemian waxwing may stay a portion of the winter or the entire winter and may breed in the area.

According to an elder, the old people before him said that when a lot of waxwings are seen together, it means that there will be a war. He adds that the bird's Dena'ina name refers to war. The name "war dancer" may refer to the bird's crest which resembles a feather headdress used by dancers, shamans, and possibly warriors.

Finches

WHITE-WINGED CROSSBILL (*LOXIA LEUCOPTERA*)

k'enchix duggets'a "twisted nose"

Description: Crossbills have a heavy bill that is crossed at the tip and white wing bars. Males are rosy red with black wings and tail. Females and juveniles are streaked brown and buff. People observe that the white-winged crossbill may stay the entire winter or only part of the winter in the Lime Village area.

According to traditional belief, a white-winged crossbill is an unlucky sign (J. Kari 1977, 235).

NOMADICS

Common Redpoll *(Carduelis flammea)*

gelujan "? branch day"

Description: A small streaked finch with a stubby bill, black chin, yellow bill, and red cap on the forehead. The adult male has a reddish breast, females and juveniles do not. Redpolls feed primarily on birch seeds and seeds of other plants. Redpolls are often abundant in an area when the birch-seed crop is good.

People report that the redpolls sometimes stay in the area only part of the winter and return in the spring while in other years, they stay throughout the winter.

Finding a redpoll nest is said to be unlucky.

Pine Siskin *(Carduelis pinus)*

generally recognized but not recorded

Description: A small streaked brown finch with a sharp pointed but short bill and bright yellow wing and tail patches visible in flight. Siskins feed on birch and alder seeds as well as seeds of herbaceous plants. The siskin has been observed during the winter in the Lime Village area.

Probable Species

The following is a list of other species that may be in the area, some of which have been reported by ornithologists visiting the area but that could not be verified by the Lime Village people.

SWAINSON'S HAWK *(BUTEO SWAINSONI)*

Description: Soars over open country with relatively long pointed wings held in a shallow *V*. The back is brown and the chest dark brown. The belly may be finely barred with orange-brown or almost clear white. The tail is barred. Some individuals may be very dark and could be mistaken for the Harlan's subspecies of the red-tailed hawk. It has not been positively identified in the area.

WESTERN WOOD-PEWEE *(CONTOPUS SORDIDULUS)*

Description: A small dusky flycatcher lacking the olive-green of the alder flycatcher and the white flank patches of the larger olive-sided flycatcher. Its call of "peeeer" and three-noted song of "tsweet-teeet-teet" is unlike the song of the other flycatchers. Wood-pewees are birds of mixed forest and although more common in eastern Alaska, could be in the Stony River drainage. This species appears as an intermediate between the more common olive-sided flycatcher and the smaller Hammond's flycatcher not mentioned by the Lime Village people, but probably in the area. It would be more likely to find Hammond's flycatcher there than the Western wood-pewee although both are possible.

HAMMOND'S FLYCATCHER *(EMPIDONAX HAMMONDII)*

Description: This small flycatcher is very hard to distinguish from the alder flycatcher mentioned earlier, and it is likely that both species are considered as one by the Lime Village people.

Hammond's flycatcher has a tiny dark bill compared to the larger and lighter bill of the alder flycatcher. Both are grayish-green above and gray to yellowish-white below depending on time of the year. Both have a thin eye ring and two light wing bars. The call of the Hammond's however is much different from that of the alder flycatcher. While the alder flycatcher sings a burry *ree-bee-oh* or *wee-bee-o*, the Hammond's sings a series of two-syllabled scratchy notes as *tsee-bik*. Hammond's flycatcher prefers drier spruce woods while alder flycatcher prefers damper areas.

SAY'S PHOEBE (*SAYORNIS SAYA*)

Description: A medium-sized gray brown flycatcher with a dark tail and rusty orange on the belly and under the tail. Say's phoebe prefers open areas such as cliffs, beaches, fields, clearings, and alpine tundra.

This species is probably at higher elevations along cliff faces in the Stony River area.

GRAY-HEADED CHICKADEE (*POECILE CINCTA*)
(formerly named Siberian tit)

Description: Similar to the boreal chickadee, the gray-headed chickadee is slightly larger and paler. The sides of the neck are white, not gray as in the boreal chickadee. Prefers willow brush and woods edge often at higher elevations. It is very hard to distinguish from the more common boreal chickadee and would be very rare in the Stony River drainage.

Townsend's Solitaire (*Myadestes townsendi*)

Description: A large gray thrush with white eye-ring and white outer tail feathers. In flight, the orange wing stripe can be seen from below. Solitaires occur in the foothills of the Alaska Range and could rarely visit the Upper Stony River area. They prefer open woodlands and forest edge often near timberline.

American Pipit (*Anthus rubescens*)

Description: Pipits are buffy brown birds of open fields, shorelines, and alpine tundra. The back is uniform brown and there are dark brown streaks on the buffy breast. The tail is dark with white outer feathers. They walk with a bobbing motion. They migrate in loose flocks often in association with longspurs and horned larks and are probably common breeders in the alpine tundra near Lime Village.

Song Sparrow (*Melospiza melodia*)

Description: A large dark sparrow usually found close to the coast and salt water in Alaska, it has a heavily streaked throat and breast with a dark central spot. Because of similarity among the streaked sparrows, reports of this species in the Lime Village area require further verification.

GRAY-CROWNED ROSY-FINCH (*LEUCOSTICTE TEPHROCOTIS*)

Description: Rosy-finches breed in alpine tundra and could be in the mountains above the Stony River. The crown is gray with a black face and yellow bill. The back and breast are brown but the belly, rump, and parts of the wing are rosy pink. They migrate to the coast in fall.

HOARY REDPOLL (*CARDUELIS HORNEMANNI*)

Description: Similar in size and general appearance to the common redpoll, hoary redpolls are lighter and have a stubbier bill. The rump and undertail feathers of the hoary redpoll are unstreaked while those of the common redpoll are lightly streaked with brown or black. Males have a rosy-pink wash on the breast and rump. Hoary redpolls are more common in the northern part of Alaska but a few could easily occur with common redpolls in winter in the Lime Village area.

Unidentified Birds with Dena'ina Names

A Snipe-like Bird, "Spanish snipe"

łiq'a ghuk'ełkeła (NL)
"the one that feeds salmon" possibly refers to the red knot, a shorebird.

A Snipe-like Bird

ts'elveni k'qilani
"the one that is like wormwood (*Artemesia telesii*)"

APPENDICES

CHickadee

Appendix A:
Lime Village
Student's Stories

This section consists of a portion of a book prepared by the Lime Village students entitled *Lime Village Birds* (Lime Village Students 1992). It was part of bird science project initiated by the teacher Melinda Moore that included a variety of bird-related activities.

The primary author was doing fieldwork for this book in Lime Village at the beginning of the Lime Village School's bird project.

> Together the students, teacher, and I inspired and helped one another with our research. A variety of activities grew from the research that the students were doing, one of which was a questionnaire about birds. Each student's questionnaire was to be included in this appendix before a computer mishap occurred that made it impossible to include them.

> In spite of the loss, the students and teacher did not let it dampen their enthusiasm. At the end of the school year, when it can be very difficult to work, they produced *Lime Village Birds;* from that the following excerpts are taken. Unfortunately it was impossible to include the entire book. The students' enthusiasm and refusal to quit in spite of a major problem, to me represents qualities possessed by the Lime Village Dena'ina as

a people. The fact that the publication was completed also seems to reflect the importance of birds to the people of the area.

Lime Village Birds

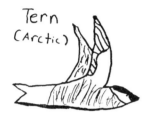

ARCTIC TERN

The Arctic Tern comes up in the spring. It goes to the end of South America in the winter because it is warm. It flies fast. He eats fish.

—Moxie Graham

GROUSE

I like to shoot grouse for food. They are good to eat. I like to hunt them all the time.

—Wasilie (Gosh) Bobby

BALD EAGLE

It has my favorite colors. It has its nest in a tree. It eats fish and meat. It lays eggs. It has nice feathers. It has no teeth. It can fly. It is big.

—Alice Willis

BALD EAGLE

It can fly. It can run in sand. It can play. They look like they have a white face.

—Briana Bobby

SWANS

Tall white and black.

They are making soft sounds in the air.

They are very proud and a brave thing about where he is going.

Feeling breezy.

—Charlie Gusty

Tundra swan

One snowy morning in September, Sue the Swan put on her shoes and ran to the school.

—Alice Willis

CHICKADEE

A chickadee can fly. It has gray feathers. It eats seeds and insects.

—Jesse Bobby

CHickadee

YELLOWLEGS (TATTLE TALE BIRD)

A tattle tale bird tells on you when you are hunting ducks. When someone is sneaking up on ducks and is trying to catch them, the tattle tale bird sits on the top of a spruce tree and tells the ducks you are around.

—Isabel Graham

LOON

A loon loves letters from a lion. He likes to laugh loudly. He likes to learn. He looks at letters, laughing books, and large leaves. He eats lettuce, lemons, and leaves for lunch.

—Evan Bobby

GOOSE

Giggle the goose gambled a game and gobbled garbage.

—Isabel Graham

OSPREY

The old osprey likes to go over the open ocean often to talk to the owl at one o'clock. The odd owl likes to talk to the osprey. When the osprey is talking to the octopus, the octopus talks to the oysters.

—Charlie Gusty

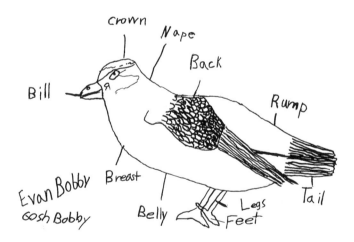

Appendix B:
Dena'ina Language Bird List

(for four dialects) with discussion (Kari, 1994)

Dialect abbreviations:

no abbreviation	=	word found in all dialects
(I)	=	Inland (Nondalton, Lime Village, often Iliamna)
(L)	=	Lime Village
(N)	=	Nondalton
(Il)	=	Iliamna
(U)	=	Upper Cook Inlet (Tyonek, Susitna Station, Knik, Eklutna)
(E)	=	Eklutna
(T)	=	Talkeetna
(O)	=	Outer Cook Inlet (Kenai, Kustatan)
(Sl)	=	Seldovia

Other symbols and abbreviations:

'. . .'	=	literal translation in single quotes
". . ."	=	local English name in double quotes
< Rus.	=	a loan word, usually from Russian
√	=	a basic root, a significant indicator of ancient form and meaning
?	=	a symbol to indicate that something is uncertain about a translation

General Terms

bird	*ggaggashla* (OU)	'little creature'
	ggaggashla gguya (Il)	
	ch'ggaggashla (I)	
a flock of birds	*nunudeti*	
a flying bird	*nunujehi*	
duck, waterfowl	*jija* (I)	√
	datishla (OU)	'little fliers'
	ye'uh nunudeti (I)	'those that fly back outside'
migrating birds	*ndati*	'plural fliers'
migrating summer birds, passerines	*shandati*	'summer fliers'
nestlings, young birds before they fly	*dedayeshghalyayi* (U)	'elevated fuzzy objects'
summer birds	*shan ch'ggaggashla* (I)	'summer birds'
	shan ggaggashla (OU)	
winter birds	*hey ch'ggaggashla* (I)	'winter birds'
	hey ggaggashla (OU)	

Birds

(Listed with some Dena'ina classification categories)

common loon *(Gavia immer)*	*dujeni* (IU) *dujemi* (O)	(possibly refers to its call)
red-throated loon *(Gavia stellata)*	*shdutvuyi* (I) *quk'ełdeti* (Ty)	possibly 'gray bill' 'the one that brings things up'
arctic loon *(Gavia pacifica)*	*ggulchun* (IO) *qulchun* (U) *ts'elba* (T) *ggagga dik* (L)	'helpless runner' 'gray one' 'creature +'
red-necked grebe *(Podiceps grisegena)*	*senya* (U) *taqa'a* (NL)	'?' 'water foot'
horned grebe *(Podiceps auritus)*	*tighuk'bet'* (T) *nachandghelahi* (I)	'dog hair stomach' 'the one that catches our scent'
red-necked phalarope, "mudsucker" *(Phalaropus lobatus)*	*hditghatl'a* (IlO) *datghatl'a* (O) *tutl'ila* (U) *taqa'a vekela* (L)	'the one that causes calm' 'water rope' 'grebe's younger brother'
swan, tundra swan *(Cygnus columbianus)*	*tava* (IU) *quggesh* (OU)	'water-gray' onomatopoetic
trumpeter swan *(Cygnus buccinator)*	*dult'iya* (IU) *nult'iya* (T) *kiłqa dudedli* (U) *tsitut'aq'a* (U)	'the one that calls *t'iy*' 'the boys are singing' 'bare head'
cormorant, double- crested cormorant, "shag" *(Phalacrocorax auritus)*	*yeq* (IO) *tsaltsiggi* (U)	√ 'the one that stuffs itself'
mature cormorant	*yeq cheh* (I)	'big *yeq*'
goose (any); Canada goose *(Branta canadensis)*	*nut'aq'i* (UO) *ndalvay* (NL) *dalvaya* (Il) *ventl'u ch'anlch'eli* (I)	'the one that flies back' 'the one that is gray-spotted' 'the one with light-colored cheeks'
black brant *(Branta bernicla)*	*chulyin viy'a* (NIl)	'raven's son'

greater white-fronted goose, "speckled-belly goose" *(Anser albifrons)*	*ndalbay* (OU) *k'dut'aq'a* (I)	'the one that is gray-spotted' possibly 'bib' or 'chest'
snow goose *(Chen caerulescens)*	*ch'iluzhena* (IO) *ch'elzheni* (I) *ch'iluna, ch'enluyna* (U)	'black wing'
ducks (any)	*jija* (I) *datishla* (UO)	√ 'little fliers'
mallard *(Anas platyrhynchos)*	*qadelchigi* (U) *chadutl'ech'i* (O) *chadatl'ech'i* (I)	'the one with yellow feet' 'the one with a blue head'
northern pintail *(Anas acuta)*	*chendghinlggesh* (I) *kadi nasa* (OU) *kadghiłnazi* (Il)	'it walks down toward the water' 'long tail'
American wigeon, "whistler" *(Anas americana)*	*ben datishla* (U) *sheshinya* (IU) *tutnełdeda* (T)	'lake duck' (name refers to a vocal sound) 'the one that chatters on the water'
northern shoveler, "spoon bill" *(Anas clypeata)*	*duyeshtala* (U) *duzhizha dghiłtali* (O) *veduzhizha dghiłtali* (I) *vedushqula* (I)	'the one whose bill is wide' 'its bill (is a) spoon'
green-winged teal, "pocket duck" *(Anas crecca)*	*qulchixa* (I) *qutnelzexa* (O) *qutnelyesha* (U)	'the one that bounces up' 'the one that flies up'
canvasback *(Aythya valisineria)*	*veq'es dasdeli* (I)	'red neck'
greater scaup, lesser scaup "bluebill" *(Aythya marila &* A. affinis)*	*jija vek'ilggeyi* (NL) *vech'enlna* *q'enk'elggeyi* (NL) *naltseghi* (Il)	'duck with white' 'the one with white on its wings' 'yellow eye'
goldeneye, common & Barrow's, "whistle wings" *(Bucephala clangula & B. islandica)*	*tsiq'unya* (NLU)	'ridged head'
bufflehead, "butterball" *(Bucephala albeola)*	*bentl'u qelch'eli* (OU) *tajehi* (U) *tsilbesi* (O) *sukna tsighał* (I)	'white cheeks' 'water puncher' 'round head' 'fabric hair bun'

long-tailed duck, oldsquaw, *(Clangula hyemalis)*	*ahhanya* (UI) *ahhangeq* (Il)	(name refers to its vocal sound)
harlequin duck, "stone duck" *(Histrionicus histrionicus)*	*qeshqa betsa'a* (OU) *denyi hdatishla* (U) *tus qet'ay* (NL)	'chief's daughter' 'canyon duck' 'resident of the passes'
common eider *(Somateria mollissima)*	*qaniłghach* (OIl) *qaniłqats'i* (U)	possibly 'the one that bites against a place'
scoter (any), "black duck"	*jijalt'esha* (I)	'black duck'
black scoter *(Melanitta nigra)*	*ułkesa qilt'ani* (U) *quk'ełdełi* (OU) *venchix va'idetsiggi* (I)	'that which looks like a bag' 'the one that brings things up' 'the one with a yellow nose'
white-winged scoter *(Melanitta fusca)*	*venaq'a qa'ilch'eli* (I)	'the one with light-colored eyes'
surf scoter *(Melanitta perspicillata)*	*venchix va'ilch'eli* (I) *veduzhizha dasdeli* (Il)	'the one with a light-colored nose' 'the one with a red beak'
common merganser, "fish duck" *(Mergus merganser)*	*cheghesh*	√
red-breasted merganser *(Mergus serrator)*	*yucheghesh* (I)	'sky + merganser'
osprey, "fish hawk" *(Pandion haliaetus)*	*tahch'ek'an* (I) *tahts'ek'e'ana* (U) *k'iltl'esa* (O)	'the one that watches the water' 'the one that seizes things'
goshawk *(Accipiter velox)*	*gizha kegh* (IO) *k'embek* (U) *k'tsu* (U)	'large gray jay' '?' √
soaring hawks, *Buteo* sp.	*q'uluq'eya* (I)	'the one that soars around'
red-tailed hawk, Harlan's hawk, spruce buteo *(Buteo jamaicensis & B. j. harlani)*	*ch'vala q'uluq'eya* (NL)	'spruce soarer'
rough-legged hawk, mountain buteo *(Buteo lagopus)*	*dghiliq' q'uluq'eya* (NL)	'mountain soarer'

golden eagle, "mountain eagle" *(Aquila chrysaetos)*	*yudi*	√
bald eagle *(Haliaeetus leucocephalus)*	*datika a* (OU) *ndatika'a* (I)	'big fliers'
marsh hawk, northern harrier *(Circus cyaneus)*	*quneh* (U) *k'kakenk'detkidza* (I)	'thrown upward' 'the one with a pattern on the base of its tail'
peregrine falcon, "duck hawk" *(Falco peregrinus)*	*tsanenh q'uluq'eya* (NL)	'cliff soarer'
gyrfalcon, "bullet hawk" *(Falco rusticolus)*	*qenay, qennay* (I)	'?'
merlin *(Falco columbarius)*	*k'enchix t'it'a* (I) *taqelquggi* (E)	'the one with a nose' '?'
northern shrike *(Lanius excubitor)*	*nak'nghitl'isha* (U) *k'eghtnitl'ishi* (O) *gizhavay* (I) *k'ededghuna* (U)	'the one that seizes' 'gray camprobber' 'warrior, killer'
owl (any); great horned owl *(Bubo virginianus)*	*besini* (U) *besi* (OIl) *mesi* (O) *k'ijeghi* (I)	√ 'the eared one'
horned owl's calls	*k'ditkidi* (Il) *k'idiki htunit* (Il)	'food' 'it'll be too much'
owl (nickname)	*tl'aq' t'etniya* (U)	'the one that makes noise at night'
snowy owl *(Nyctea scandiaca)*	*yesvu*	'snow white'
northern hawk owl *(Surnia ulula)*	*delukdiday* (I) *duh besina* (T) *luk'chen* (U)	'the one that sits on branches' 'timberline owl' '?'
great gray owl *(Strix nebulosa)*	*nutdes*	'?'
short-eared owl *(Asio flammeus)*	*naghuk'ts'eha* (OU) *naghuk'etts'eha* (I)	'the one licking something for us'

boreal owl *(Aegolius funereus)*	*skindezduya* (IL) *skindeyduya* (U) *skitnaz'una* (NL) *skimbaya* (U)	'the one that stays under trees' 'the white one under trees'
spruce grouse, "spruce chicken" *(Dendragapus canadensis)*	*ełyin* (OIl) *ełdyin* (NL) *ełyuni* (U)	'the one that eats spruce boughs'
ruffed grouse, "willow chicken" *(Bonasa umbellus)*	*k'dełneni* (IO) *chugget'a* (U)	'the one that pounds' '? striker'
sharp-tailed grouse *(Tympanuchus phasianellus)*	*k'etteli*	'the one that makes a thumping noise'
ptarmigan (any); willow ptarmigan *(Lagopus lagopus)*	*delggema* (OU) *q'ach'ema* (I)	'that which calls *ggem*' '?'
rock ptarmigan *(Lagopus mutus)*	*q'ach'ema* (U) *jeł q'ach'ema* (NL) *dghili q'ach'ema* (Il)	'?' 'mountain ?' 'mountain ?'
white-tailed ptarmigan *(Lagopus leucurus)*	*qatsinłggat* (I) *ch'etl' q'ach'ema* (I) *dzeł yicheghi* (U) *dus yicheghi* (O)	'you are dreaming' 'willow white' 'mountain crier' 'distant crier'
black oystercatcher *(Haematopus bachmani)*	*tiq'a chihi* (Il)	'it cries in the mud flats'
sandhill crane *(Grus canadensis)*	*ggukdel* (O) *ndał* (UI), *nedał* (Il) *k'q'eshch'a ghe'uti* (T)	'*gguk* (noise)-red' 'plural fly' 'the one that chews its jowls'
golden-plover, American & Pacific *(Pluvialis dominica & P. fulva)*	*ggulyit*	'speed chaser'
semipalmated plover *(Charadrius semipalmatus)*	*sank'tnal'ay* (NL) *sank'tnal'iy* (O) *talyiya* (U)	'the one with a ring around its collar' '?'
surfbird *(Aphriza virgata)*	*yudi ghelkala* (I)	possibly 'golden eagle tail'

Wilson's snipe, "jack snipe" (*Gallinago delicata*)	*yuził* (OI) *yuyił* (U)	'sky whistler'
whimbrel *(Numenius phaeopus)*	*duzhish dghulggesha* (L) *nuduyesdghulggesha* (I) *veduzhizha dghelggesha* (Il)	'its beak is curved'
yellowlegs: lesser and greater, "tattletale bird" *(Tringa melanoleuca & T. flavipes)*	*sadiya* (O) *sudiya, sudya* (U) *sadya* (I)	(name refers to its vocal sound)
spotted sandpiper *(Actitis macularia)*	*delvizha* (I) *tabagh telggesha* (U)	'the one that calls *vizh*' 'the one that runs on the beach'
small sandpipers; semipalmated, Western, and least sandpipers *(Calidris pusillus, C. mauri, C. minutilla)*	*qenghesh k'ela* (U) *qenghish k'ela* (I)	'foam tearer'
dowitcher: short-billed & long-billed *(Limnodromus griseus & L. scolopaceus)*	*kadantsa* (I)	'tail ?'
long-tailed jaeger, "shit gull" *(Stercorarius longicaudus)*	*nuk'ełvaq'i*	'the one that induces vomiting'
gull (any)	*nulbay* (U) *vach* (OI)	'the one that is gray' √
large gulls, glaucous-winged & glaucous *(Larus glaucescens & L. hyperborus)*	*vach kegh*	'large gull'
herring gull *(Larus argentatus)*	*łiq'a vaja* (I) *tl'iq'a beja* (U)	'salmon's gull' 'mud flat's gull'
mew gull *(Larus canus)*	*shagela vaja* (I)	'trout's gull'
Bonaparte's gull *(Larus philadelphia)*	*chilzhena* (IO) *tsilyeni* (U)	'the one that has a black head'

black-legged kittiwake *(Rissa tridactyla)*	*gedeyaq* (U)	(name refers to its vocal sound)
	gadiyaq (O) *gadayaq* (Il)	
arctic tern *(Sterna paradisaea)*	*ts'ik'nal'uya* (O) *ts'ik'nal'i* (U) *ch'ink'nul'ay* (NL) *ch'ik'enal'uya* (Il) *shtlaq* (L)	'it keeps its head erect' '?'
puffin: horned & tufted *(Fratercula corniculata, F. cirrhata)*	*duzhizha delchezhi* (IlN) *duyiya delcheyi* (U) *tiq'ats'i* (O)	'its beak is a rattle' 'water ?'
common murre *(Uria aalge)*	*shangideq* (O)	< Alutiiq
belted kingfisher *(Ceryle alcyon)*	*chik'dghesh* *tiq'a qeshqa* (T) *tiq'a ten nuqughuni* (U)	'ragged head' 'salmon's chief' 'the one the makes the salmon's trail'
northern flicker *(Colaptus auratus)*	*tsenet* (UO)	'nose-wedge'
woodpecker (any)	*tsik'tnetqetl'a* (U) *k'detggetla* (O) *kuntsulya* (I)	'the one that chops with its head' 'the one that knocks' 'stomach + ?'
hairy woodpeckers, alder or cottonwood woodpeckers *(Picoides villosus)*	*qenq'eya kuntsulya,* *eseni kuntsulya* (I)	'alder woodpecker,' 'cottonwood woodpecker'
downy woodpecker, willow woodpeckers *(Picoides pubescens)*	*ch'etl' kuntsulya* (I)	'brush, willow woodpecker'
northern three-toed and black-backed woodpeckers, spruce woodpeckers *(Picoides tridactylis & P. arcticus)*	*ch'vala kuntsulya* (I)	'spruce woodpecker'
red-breasted nuthatch *(Sitta canadensis)*	*kenghutelggesha* (U) *kuntsulya gguya* (L) *kenegh itghichesha* (N)	'the one that goes along flat' 'little woodpecker' 'the one that hangs upside down'
brown creeper *(Certhia familiaris)*	*kuntsulya gguya* (L)	'little woodpecker'

gray jay, "camprobber"	*taqelbi* (U)	'? gray'
(Perisoreus canadensis)	*ch'k'naqałch'eya* (I)	'?'
	k'naqałch'eya (Il)	
	gizha (O)	√
Steller's jay, "blue jay"	*ułchena gizha* (O)	'Alutiiq camprobber';
(Cyanocitta stelleri)	*ułchena ggaggashla* (U)	'Alutiiq bird'
magpie *(Pica*	*k'ahtal'iy* (U)	'the proud one'
hudsonia)	*q'ahtghal'ay* (NL)	
	k'ahtal'aya (Il)	
	k'ahtal'uya (O)	
	shehutniga (T)	'?'
	chuqutniga (L)	
common raven	*chulyin* (I)	'the one that eats shit'
(Corvus corax)	*ggugguyni* (O)	'the creature'
	delgga (U)	'the one that caws'
northwestern crow	*chinshla* (OIl)	'little shit'
(Corvus caurinus)		
chickadee (any);	*naghuynisdi* (U)	'our thoughts'
boreal chickadee	*ch'ggagga* (I)	'creature'
(Poecile hudsonica)	*ch'qeshniha* (O)	'the one that says *ch'qesh*'
black-capped chickadee	*tsidut'aq'a* (I)	'bare headed'
(Poecile atricapilla)		
olive-sided flycatcher	*dehvava, vava nihi* (I)	'their dry fish';
(Contopus cooperi)		'one that says dry fish'
	shdaja (U)	'our little sister'
	k'bet' didi (T)	'it says stomach'
its call	*benuja deldeli dnił'ił* (U)	'you'll see red fish meat'
swallow, "tom swallow"	*tl'ałghak'a* (NL)	'firedrill ?' or 'flint ?'
"blue back" *(Tachycineta*		
bicolor)		
swallow (any); violet-	*katja* (NL)	'?'
green swallow	*kuntjeja, kunjeja* (U)	
(Tachycineta thalassina)	*kunjija* (O)	
	kuntjeja (Il)	
bank swallow	*vest'ugh katja* (NL)	'beneath the bank swallow'
(Riparia riparia)		
cliff swallow	*qenk'denghełts'etl'i katja,*	'swallow that daubs
		something'
(Hirundo pyrrhonota)	*tsanenh katja* (NL)	'cliff swallow'

American dipper, water ouzel (*Cinclus mexicanus*)	*tudzelqet'a* (U) *tatsilt'ana* (O) *tatsilqit'a* (I)	'the one that stays at open water' 'the one who has its head in water'
American robin (*Turdus migratorius*)	*shih* (U) *kałnay* (IO) *grasnay bushga* (O)	√ '?' < Rus.
varied thrush (*Ixoreus naevius*)	*deltl'ishi* (OI) *deltl'ezhi* (Il) *skintu duhdnghisesi, dkentu k'enisesa* (U) *dli ni* (T)	'the one that calls *tl'ish*' 'the one that echoes through the trees' 'the one that says cold'
hermit thrush (*Catharus guttatus*)	*kentuchila* (I) *ketuchalashi* (N) *luk'delghuzha* (O)	'meadow-water-moist' 'the one that sings in the branches'
Swainson's thrush, "big-eyed bird" (*Catharus ustulatus*)	*shinggach'* (I)	'it is staring at me'
gray-cheeked thrush (*Catharus minimus*)	*ggezhaq* (I) *eyah niyi* (U)	(refers to its vocal call) 'it says filthy'
ruby-crowned kinglet (*Regulus calendula*)	*deghgich* (I) *jatnuldeli* (O)	'hanging cartilage' or 'hanging rustling noise' 'one with a stripe of red'
yellow warblers: arctic, orange-crowned, yellow & Wilson's (*Phylloscopus borealis, Vermivora celata, Dendroica petechia, & Wilsonia pusilla*)	*ltseghay, k'eltseghay* (I) *k't'unltsegha* (U)	'yellow one' 'the one with yellow feathers'
yellow-rumped warbler (*Dendroica coronata*)	*k'densuya* (I)	'the one that makes rattling noise'
northern waterthrush (*Seiurus noveboracensis*)	*nent'ughazh* (I) *kala k'detts'etl'i* (U)	'under the ground round object' 'its sticks with its tail'
Bohemian waxwing (*Bombycilla garrulus*)	*k'eghunjija, k'eghunjech'a* (O)	'war dancer'
rusty blackbird (*Euphagus carolinus*)	*ch'qełch'ah* (IU) *nudujeła* (O)	(refers to its vocal call) 'the one that hollers over and over'

pine grosbeak *(Pinicola enucleator)*	*hey dudiya* (U) *hidudya* (O) *qidudya* (I) *k'qidudya* (Il)	'winter -?' (refers to its vocal call)
common redpoll *(Carduelis flammea)*	*din nihi* (U) *gelujan* (I) *k'elujan* (O)	'the one that says *din*' '-? day'
white-winged crossbill *(Loxia leucoptera)*	*k'enchix duggets'a* (I) *chish duggets'a* (U)	'twisted nose'
Lincoln's sparrow *(Melospiza lincolnii)*	*tl'ujech'a* (L)	'tail dancer'
dark-eyed junco *(Junco hyemalis)*	*k'dett'edi* (I) *delt'ech'a* (OU) *veka ghundalggeyi* (N) *daljeja* (U)	'the one that calls *t'et* (smacking sound) 'the one that calls *t'ech*' 'its tail is part white' 'the one that whispers'
white-crowned sparrow *(Zonotrichia leucophrys)*	*ggulush didi* (I)	'?'
golden-crowned sparrow *(Zonotrichia atricapilla)* its songs	*tsik'ezdlagh* (OI) *tsisk'eydlan* (U) *shishk'ezet'a* (Il) *nulchina nutgheshjuł* (U) *chik'a dulnił* (I)	'it is *tsi*' 'it is *tsi, tsis*' 'my liver' 'I'll return to the Sky Clan' 'wood will be'
fox sparrow *(Passerella iliaca)*	*dushidaghiłchiq'i* (I)	'why does it scold me?'
savannah sparrow *(Passerculus sandwichensis)*	*ch'ich' qunsha* (I)	'ground squirrel that goes *ch'ich'* (scraping noise)'
American tree sparrow *(Spizella arborea)*	*ch'nutch'ish* (NL)	(refers to a rustling, scraping noise)
snow bunting, "snow bird" *(Plectrophenax nivalis)*	*hggush* (U) *ggush* (O) *nehggux* (NL) *hggux gguya* (Il)	'grayish'
unidentified bird	*ts'elveni k'qilani* (N)	'the one that is like wormwood (plant)'
unidentified bird	*łiq'a ghuk'etkeła* (NL)	'the one that feeds salmon'

Domestic or Distant Birds

chicken	*gulitsa* (UO) *gulutsa* (I)	< Rus.
rooster	*bidux* (O)	< Rus.
peacock	*hudelkuhi ggaggashla* (O)	'vain bird'
ostrich	*k'kegh'i ggaggashla* (O)	'giant bird'

Parts of Birds

beak	*k'duzhuzha* (O) *k'edudzhizha* (I) *k'duyiya* (U)	
gizzard	*k'kuluzhuma* (O) *k'jeha* (IU)	√
breast meat	*k'yaytsen* (U) *k'eyaɫtsen* (I) *k'eyatsen* (O)	
breastbone	*k'bayqa* (U) *k'eyaɫqa'a* (IO)	'its canoe'
crop sack in grouse's or ptarmigan's neck	*k'geɫq'aɫtuna* (OU) *k'geɫ q'aɫtana* (I) *k'ggaɫgguts'a* (O)	
claw	*k'qaggena*	
talon	*k'qajuga*	
egg	*k'eghazha* (OI) *k'eghaya* (U)	
eggshell	*k'ghazha ghes* (O) *k'ghayts'ena* (U) *k'ghazhts'ena* (I)	
egg yolk	*k'ghazh chiga* (I) *k'ghay chiga* (U) *belchin'a* (O)	

tail	*k'ka*	√
wing	*k'ch'elna* (O) *k'ch'enla* (Il) *k'ch'enlna* (NL)	
base of feathers, bumps on skin	*k't'ukena*	
long feathers, wing or tail feather, quill	*k't'u*	√
wing feathers	*k'ch'enlna t'u*	
tail feathers	*k'ka t'u*	
plucked medium longish feathers or fur	*k'kidza*	
attached strands of fur or long feathers	*k'andaz'i*	
fine down feathers	*k'keshch'a, kets* (U) *ditushi* (OUIl) *dituxi* (NL)	
crown of feathers	*betsik'et'a* (U)	
tufted head feathers (on waxwing, kingfisher)	*k'ank'dasdlits'i* *qadadlits'a* *vetsiduq' k'andazdlits'i* (I)	
nest	*k'det'eh, k't'eh*	
ground nest	*k'eghazhq'a* (IO) *k'eghayq'a* (U)	'egg cavity'
mud nest	*qenk'denghetts'etl'i* (IU)	'that which is daubed on a place'

Discussion (by James Kari)

This presentation of bird names for the whole Dena'ina language area provides an opportunity to make a number of generalizations about (1) the interrelations between the Dena'ina dialects, and (2) the nature of research on bird names in a language with small communities of speakers, some of which are now not spoken by many persons.

The Dena'ina language is geographically very diverse, containing the marine environment of Cook Inlet and being on both sides of the southern Alaska Range. A dictionary of the Dena'ina language, such as J. Kari (1977), has many regional subpatterns for words of the same meanings.

The Dena'ina language now has the best documentation on birds across dialects of any Alaska Native language. With 120 identified birds across the large language area, this is the most detailed and best organized multi-dialect bird list, eclipsing the Koyukon bird list with 98 identified birds (Jettè and Jones 2000) and probably also Aleut (Bergsland 1994, where the data have not been consolidated.) The coverage for Lime Village is unique in that it has extra attention to identification issues such as the lumping of closely related species with a single term and the recognition of another 18

SUMMARY OF DISTRIBUTIONAL PATTERNS ON BIRD NAMES IN THE DENA'INA DIALECTS

a. total number of identified birds (species or cluster of species)

all of Dena'ina	Inland	Lime	Upper	Outer
120	116	111	83	70

b. Lime Village 111 names for 127 species of birds and 18 other birds recognized but not named

c. identified birds restricted to certain dialect areas either by habitat or by gaps in dialect information.

Inland only	Inland+ Upper	Upper only	Outer- Upper	Outer only	Iliamna- Outer
37	18	1	6	2	3

d. distinct terms for 120 birds throughout Dena'ina dialects: 210

e. birds documented/known for all dialects: 59

f. birds with a cognate terms in all dialects: 25

g. bird names marked by a unique root or stem (√): 10

birds which do not have Dena'ina names. Note also that our coverage of the birds is not even across the language. For Kenai and the Outer Inlet dialect we mainly have names for the more common and well-known birds. We have more expert-level information for Upper Inlet, and we have truly fine-grained identifications and terminology for the Inland dialect (Lime Village and Nondalton). The Lime Village-Nondalton ornithological subtleties such as subcategories for hawks, woodpeckers, and the recognition of most of the summer song birds, are based upon acute observations and are a wonderful display of folk science.

As we examine the bird names across the dialects, we find a great diversity in names, with about 210 distinct names for 120 birds. This figure does not count the pronunciation differences in names that are cognate. For example, for violet-green swallow there are five variant forms in the dialects, but these are counted as one distinct name term. For magpie there are two distinct terms, one of which has three variant pronunciations: *k'ahtal'iy* (U), *q'ahtghal'ay* (I), *k'ahtal'uya* (O), and one of which has two: *shehutniga* (T) and *chuqutniga* (L). These dialect patterns are interesting. We see at times that all dialects have a common term, for example *yuzit* or *yuyit* Wilson's snipe, or two or three of the dialects share one name and one has a distinct name such as *vach* (IO) and *nulbay* (U) for gull. Some of the Dena'ina dialect patterns are governed by a bird's habitat. Only the Outer Inlet dialect has names for marine birds such as black oystercatcher, northwestern crow, and murre. Horned puffins were known across three dialects because their beaks were used in ceremonial rattles.

Some Dena'ina bird names are found elsewhere in the Athabascan language family, another level of patterning which Appendix B does not attempt to display. For example, both of the terms for gull mentioned above are found in other languages. Cognates with *nulbay* are in Ahtna and Deg Hit'an and cognates with *vach* are more widespread in most of Alaska Athabascan and in some Canadian Athabascan languages. Terms related to *ndat* sandhill crane are found in many languages including Carrier in British Columbia and Navajo in the Southwest.

Appendix C:
References

American Ornithologists' Union. 1998. *Check-list of North American Birds*, 7th ed. Washington, D.C. liv + 829pp.

Bergsland, K. 1994. *Aleut Dictionary*. Fairbanks: Alaska Native Language Center.

Bureau of the Census. 1992. *1990 Census of Population and Housing, Summary: Social, Economics, and Housing Characteristics, Alaska*. U. S. Printing Office, Washington, DC.

Jetté, J., and E. Jones. 2000. *Koyukon Athabaskan Dictionary*. Fairbanks: Alaska Native Language Center.

Kalifornsky, P. 1991. *A Dena'ina Legacy, K'tl'egh'i Sukdu: The Collected Writings of Peter Kalifornsky*. Fairbanks: Alaska Native Language Center.

Kari, J. 1974. *Kenai Tanaina Noun Dictionary*. Preliminary version. Fairbanks: Alaska Native Language Center.

———. 1977. *Dena'ina Noun Dictionary*. Fairbanks: Alaska Native Language Center.

———, ed. 1983. *K'qizaghetnu Ht'ana*. Anchorage: National Bilingual Materials Development Center.

———. 1988. Some Linguistic Insights into Dena'ina Prehistory. In *The Late Prehistoric Development of Alaska's Native People,* edited by R. D. Shaw, R. K. Harritt, and D. E. Dumond. Aurora, Alaska Anthropological Association Monograph Series 4, pp. 319–39.

———. 1994. *Dictionary of Dena'ina Athabaskan, Vol. 1: Topical Vocabulary* (draft). Fairbanks: Alaska Native Language Center.

Kari, J., and J. Fall, eds. 1987. *Shem Pete's Alaska, The Territory of the Upper Cook Inlet Dena'ina.* Fairbanks: Alaska Native Language Center.

Kari, J., and P. R. Kari. 1982. *Dena'ina Etnena, Tanaina Country.* Fairbanks: Alaska Native Language Center.

Kari, P. R. 1983. Land Use and Economy of Lime Village, Tech. Paper 80. Alaska Department of Fish and Game, Anchorage, Alaska.

———. 1987. *Tanaina Plantlore: Dena'ina K'et'una.* Anchorage: National Park Service, Alaska Region, Anchorage, AK.

Lime Village Students. 1992. Lime Village Birds. Lime Village School, Iditarod School District, Lime Village, AK.

Osgood, C. 1937. *The Ethnography of the Tanaina.* Yale University Publications in Anthropology, No. 16. New Haven: Yale University Press.

Tenenbaum, Joan M. 1978. *Nondalton Tanaina Noun Dictionary.* Fairbanks: Alaska Native Language Center.

Tenenbaum, J., and M. J. McGary, eds. 1984. *Dena'ina Sukdu'a: Traditional Stories of the Tanaina Athabaskans.* Fairbanks: Alaska Native Language Center.

Viereck, L. A., and C. T. Dyrness. 1980. A Preliminary Classification System for Vegetation of Alaska, General Tech. Report PNW–106, USDA Forest Service, Pacific Northwest Forest Range and Experiment Station, Fairbanks, AK.

Viereck, L. A. and E. L. Little, Jr. 1972. *Alaska Trees and Shrubs.* USDA Forest Service, Washington, DC.

Wassillie, A. Sr. 1980. *Huvendaltun Ht'ana Sukdu'a: Nondalton People's Stories.* Anchorage: National Bilingual Materials Development Center, University of Alaska Anchorage.

———. 1980. *K'ich'ighi: Dena'ina Riddles.* Anchorage: National Bilingual Materials Development Center.

West, G. C. 2002. *A Birder's Guide to Alaska.* Colorado Springs: American Birding Association (viii, 586pp).